TRANSACTIONS OF THE INTERNATIONAL ASTRONOMICAL UNION
VOLUME XXX

PROCEEDINGS OF THE THIRTIETH GENERAL ASSEMBLY VIENNA 2018

COVER ILLUSTRATION:

Paranal Observatory at Dusk: Dramatic dusk with dissipating thin clouds over ESO Paranal Observatory captured on 4 July 2019 (just 2 days after the total solar eclipse at La Silla). The crescent Moon can be seen on left.

Auxiliary telescopes, Cerro Paranal, VLT Unit Telescopes

Credit: ESO/P. Horalek

IAU SYMPOSIUM PROCEEDINGS SERIES

Chief Editor
MARIA TERESA LAGO, General Secretary
Universidade do Porto
Centro de Astrofísica
Rua das Estrelas
4150-762 Porto
Portugal
iau-general.secretary@iap.fr

Editor
IAN ROBSON, Assistant General Secretary
Royal Observatory Edinburgh
UKATC
Blackford Hill
Edinburgh
EH9 3HJ
UK
IAU-assistant.general.secretary@iap.fr

INTERNATIONAL ASTRONOMICAL UNION

UNION ASTRONOMIQUE INTERNATIONALE

TRANSACTIONS OF THE INTERNATIONAL ASTRONOMICAL UNION VOLUME XXX

PROCEEDINGS OF THE THIRTIETH GENERAL ASSEMBLY VIENNA 2018

Edited by

DEBRA MELOY ELMEGREEN

IAU President (2021–2024)

CAMBRIDGE
UNIVERSITY PRESS

CAMBRIDGE UNIVERSITY PRESS
University Printing House, Cambridge CB2 8BS, United Kingdom
1 Liberty Plaza, Floor 20, New York, NY 10006, USA
10 Stamford Road, Oakleigh, Melbourne 3166, Australia

First published 2024

Printed in Great Britain by Henry Ling Limited, The Dorset Press, Dorchester, DT1 1HD

Typeset in System LaTeX 2ε

A catalogue record for this book is available from the British Library of Congress Cataloguing in Publication data

This journal issue has been printed on FSC™-certified paper and cover board. FSC is an independent, non-governmental, not-for-profit organization established to promote the responsible management of the world's forests. Please see www.fsc.org for information.

ISBN 9781108488723 hardback
ISSN 1743-9213

Table of Contents

Preface

The IAU *Transactions B* series report the activities that take place during the General Assembly (GA). These reports record all the essential decisions taken by the main governing body of the IAU: the approval of the financial accounts and of the proposed budget for the next triennium, the admission of new National Members and Individual members, the evaluation of Division and Commission reports, the approval of Resolutions, etc.

The General Assembly welcomed the re-admission of Algeria, Jordan, Morocco (Observer), and the admission of Ghana, Cyprus, Madagascar (Observer), Mozambique (Observer), Syrian Arab Republic (Observer) and United Arab Emirates as a National Member. The General Assembly also welcomed the admission of 1240 new Individual and Junior Members from 41 Countries.

At the General Assembly there was the launching of the IAU 100 years celebration, which included the "Under One Sky: The IAU Centenary Symposium" and other events around the world.

The Proceedings of the 6 Scientific Symposia held during the GA and of the "Under One Sky: The IAU Centenary Symposium" that was launched during the GA were published separately by Cambridge University Press in the Series IAU Symposia.

The Proceedings of the 13 Focus Meetings that took place during the GA were published separately by Cambridge University Press in the "Astronomy in Focus" volume.

A sincere thank to the University of Vienna, the Austrian Academy of Sciences and the Austrian Society for Astronomy and Astrophysics for organising a very exciting and pleasant General Assembly, in the beautiful city of Vienna.

Teresa Lago, General Secretary (2018–2021)

Addendum:

The world was forever changed not long after the 2018 General Assembly. This XXX GA volume was delayed due to the coronavirus pandemic and other disruptions, so was prepared by the subsequent administration. Because this was done five years after the General Assembly, some information were not recoverable. Since *Transactions A* has been replaced by *Astronomy in Focus*, hereafter the GA volumes will be known as *Transactions* rather than *Transactions B*. This volume contains all of the items available as of August 2023. I am grateful to Piero Benvenuti, Silvia Torres-Peimbert, Ewine van Dishoeck, Teresa Lago, and Gerhard Hensler for their assistance in providing materials.

Debra Meloy Elmegreen, President (2021–2024)

Transactions IAU, Volume XXX
Proc. XXX IAU General Assembly, August 2018
Debra Elmegreen, ed.
doi:10.1017/S1743921323005598

CHAPTER I

THIRTIETH GENERAL ASSEMBLY

INAUGURAL CEREMONY

Monday, 20 August 2018, 16:30-18:00

Austria Center, Vienna

1. Opening Event

The XXXth General Assembly was opened by Anneliese Haika from the Local Organizing Committee and National Organizing Committee at the Austria Center, following a musical introduction by the Habe Quartett.

Welcome Addresses were given by the following dignitaries:

Alexander Van der Bellen, Federal President of the Republic of Austria (by video)

Heinz Fassmann, Austrian Federal Ministry of Education, Science and Research
Ladies and gentleman,

It is a great pleasure to welcome the General Assembly of IAU to Austria. Astronomy and astrophysics have a long history in Austria and current developments are very dynamic. This is especially due to Austria's inclusion in ESO 10 years ago. This dynamics is also true on a European and global scale. With competent and enthusiastic people all around the world and a variety of state-of-the-art instruments ranging from telescopes to satellites, an impressive amount of new knowledge and insight into the universe is gained at a really breath-taking speed.

On the day this note was written, *Nature* published a lead article about long-standing research on the massive centre of our galaxy and its impact on nearby stars. It's a fascinating piece of work as we cannot see the real thing, but measure it and its gravitational impact. This is yet another impressive result and shows what the combination of scientific theory and experiment is capable of.

I am very happy that Austria has established itself as a competent partner in the global scientific landscape and I am sure the years ahead will continue to produce many more interesting results.

Heinz Engl, Rector of the University of Vienna
Dear Colleagues, Ladies and Gentlemen,

As Rector of the University of Vienna, I welcome you to the International Astronomical Union General Assembly in Vienna. The International Astronomical Union (IAU) is the largest association of professional astronomers. It currently has 9623 members from 86 countries. And the University of Vienna is the largest University in the German speaking countries with 93000 students and around 10000 employees (6500 of them academic teachers and researchers) and 19 faculties. Astrophysics is located at a Faculty which we call informally "the Faculty of Exploration," namely the Faculty of Earth Sciences, Geography and Astronomy. The IAU was founded in 1919 in Brussels. Every 3 years, 3000–4000 IAU members convene for a "General Assembly" (GA). There have been 29 GAs since the foundation of the IAU. The present, the 30th IAU GA, is the first one ever to take place in Vienna following the ones in Prague in 2006, in Rio in 2009, in Beijing in 2012, and finally in Hawai'i in 2015. Very different places indeed, each with its own character; we hope Vienna can compete well both scientifically and as a pleasant town.

The University of Vienna was founded in 1365, 17 years after Prague. Already in the first century of its existence, astronomy (and mathematics) flourished at the University of Vienna: Johannes von Gmunden, Georg von Peuerbach, Regiomontanus were all professors here just to mention a few famous ones. Thus, Vienna was important in preparing the Copernican view of the world. And we should also remember that Johannes Kepler was active in cities not far from Vienna, namely Graz, Linz (where the current University is named after him) and finally Prague.

To jump over a few centuries, let me mention that astronomy and astrophysics is a quite active and successful department, which the University is proud of. A big boost to astrophysics, which also enabled us to make a few excellent hirings, was given by Austria joining ESO a few years ago. As a mathematician, I was involved in the negotiations and also initiated a large mathematical project on adaptive optics. Hence, I am closer to the field of your conference than at usual welcome speeches by a Rector.

Thus, I give you a special welcome, thank the IAU for holding their GA here and especially thank the local organizers under the leadership of Prof. Gerhard Hensler for all the hard work they have done and will still be doing. Have a nice stay in Vienna.

Manuel Guedel, President of the Austrian Soc. of A&A
Welcome to Austria, welcome to Vienna, and welcome to the General Assembly of the International Astronomical Union!

The Austrian Society for Astronomy and Astrophysics (OeGAA) is the proposing organization for the IAU GA 2018; we are proud of seeing a large international community of astronomers meeting in Vienna to exchange new astronomical results.

Astronomy in Austria has seen a rapid growth over the past two decades; its researchers work at the cutting edge of science, in areas such as solar physics, solar system research, exoplanets, star formation and stellar evolution, or galaxy formation and evolution. The OeGAA acts as the umbrella organization representing the interests of our scientists. A first culmination and in fact a top priority of OeGAA was the accession of Austria to ESO a decade ago, in 2008; our new membership in turn motivated universities to announce several new astronomy professorships, and the government and our universities to support the joint initiative of our astronomical institutes to contribute to three instruments for ESO's Extremely Large Telescope (ELT), two first-light instruments among them.

Austria also enjoys a long tradition in developing payload components for space missions and observatories, both for solar-system research and astrophysics. ESA is our closest partner for such projects, and national collaborations in payload development have become a standard. But apart from professional astronomers, OeGAA equally takes amateur astronomers and their associations on board. And likewise, it functions as an active link between the government, the public, academic research, and international organizations.

And so the loop closes: After our successful initiative and preparations led by our member Gerhard Hensler, we are glad to finally reach the culmination of the largest astronomical gathering this country has seen, the IAU GA 2018. I thank the IAU for a productive partnership over the past several years, and I wish you all enjoyable days in Vienna and fruitful scientific discussions!

A musical interlude followed.

2. Opening Address

Dr. Silvia Torres-Peimbert, IAU President

Dr. Heinz Fassmann, Austrian Federal Minister of Education, Science and Research, Dr. Heinz Engl, Rector of the University of Vienna, Prof. Manuel Guedel, President of the Austrian Soc. of A&A, Prof. Gerhard Hensler, Head of the National Organizing Committee, National Representatives, Dr. Heide Heckman, International Science Council, Mrs. Patricia Gruber, Gruber Foundation, IAU members and invited participants: Dear colleagues and friends, I have the honor of welcoming you to the XXX General Assembly of the International Astronomical Union. We have come to meet in Vienna in this wonderful city and its rich cultural heritage. It is inspiring to be in the same place where so many talented artists and scientists flourished and are presently active.

In this meeting where more than 3,000 participants have registered from 89 countries, we re-unite again, of course to meet old friends and make new acquaintances. Key topics in contemporary astronomy will be addressed, and the latest scientific progress in a number of specialised areas will be presented. There will be an exciting exchange of scientific knowledge. This conference will, no doubt, stimulate discussion in cutting-edge astronomy in the different fields where it is being advanced; and more importantly it will encourage: international communication, exchange of ideas and consolidate collaborations resulting in important advancements for the astronomical community and beyond.

This meeting is about an IAU which is very actively creating for itself a future relevant for its next 100 years of existence. The community of astronomers is indeed an international community. We can move with enormous freedom around the world, work in different countries, and the national barriers are not strong limitations to what we do. This is a wonderful privilege that we have in this time and in our area of research.

The program includes seven Symposia and fifteen Focus Meetings that cover a wide spectrum of contemporary astrophysics. There will be Division Meetings to facilitate participants to share their work with the rest of our colleagues, as well as Commissions and Working Groups meetings where members will continue their scientific and organizational dialogs. Please be aware that each of these activities has secured time slots for Poster sessions which we encourage you not to miss.

In addition to the Special Session on the General relativistic effects in stellar orbits around the Galactic Center black hole that has just taken place, there will be other plenary sessions where all of you are invited to attend; there will be three Invited Discourses, and the Gruber Prize talk. These plenary sessions have been designed to cover different topics of current development. Furthermore, for each Symposium its scientific organizing committee has selected plenary talks to share with us the exciting new results in its field of interest. Among the general activities there will be lunch meetings devoted to issues of interest to Young Astronomers and to Women in Astronomy.

We met 3 years ago, and in this period many changes have taken place in the organization: we will propose a set of modifications to the Statutes, Bye-laws and Working Rules that try to capture the ever-changing needs of this society. There are 10 new National Members, and about 1000 new Regular Individual Members from around the world that will join the Union. And as is customary in these meetings, a set of Resolutions will be presented to you for approval.

We have been very concerned about including younger astronomers in our organization. For this purpose, among the changes in statutes we have proposed to create a new category of membership, Junior Member, to include astronomers that have recently obtained their PhD, for a duration of up to 6 years in this class. By the way, this increment on membership will not represent an increase in dues.

Upon approval of this proposal, we can welcome 352 Junior Members to the Union. We propose to establish a Young Astronomers Working Group to address their concerns. To further encourage interest from young astronomers to the IAU we established in 2016 a new prize, called the IAU PhD Prize, to recognise outstanding scientific achievement in astrophysics around the world. Each IAU Division will, once a year, have the opportunity to award its own prize to the candidate it feels has carried out the most remarkable work in the previous year. In addition, a separate prize is available to be awarded to applicants from developing countries. So far, 19 outstanding young astronomers have been selected to receive this award.

As you all know, this General Assembly is very special, we commemorate the first 100 years of existence of the International Astronomical Union which was founded in 1919. The Union is an old organization that has renewed itself continuously and has extended its activities to preserve and increase its value to the community. To mark this important milestone one of the seven symposia is dedicated to review the history and accomplishments of the IAU. On Monday 27, following the Symposium 349 Plenary talk, a Centennial Ceremony will take place to which all of you are invited. In this building an extraordinary exhibit has been set up highlighting the last one hundred years of astronomical progress, Under one sky, Above and Beyond. This exhibit is designed to be transported to various locations over the next year, starting with the opening of the anniversary celebrations in Brussels on 11 and 12 of April. To commemorate this achievement, the IAU is organising a year-long jubilee under the central theme "IAU 100 years - Under One Sky". It includes an extensive set of activities to be carried out around the world, for the purpose of increasing awareness of a century of astronomical discoveries as well as for supporting and improving the use of astronomy as a tool for education, development and diplomacy. We invite you all to participate and extend this celebration in your home country.

Since 2009 the IAU transformed itself from an organization with a largely internal focus emphasizing meetings and events for its members to one more involved in education and outreach to the general public. To this effect the Astronomy for Development Strategic Plan for 2010–2020 approved in 2009 and extended in 2015 gave rise to the Office of Astronomy for Development in Cape Town; similarly, the Office of Astronomy Outreach in Tokyo, was established and the virtual Office for Young Astronomers that encompasses the International School of Young Astronomers was formalized in Oslo. In these three cases, we receive generous support from the host countries. On this occasion we are proposing an integral project, the Strategic Plan 2020–2030. It is not just an update of the existing "Astronomy for Development Strategic Plan 2010–2020," rather it is an attempt to promote all the activities of the Union in the next decade.

In the exhibition areas there are displays from our institutions and from the instrumentation companies that sponsor this event. Also, there is the travelling international exhibition Inspiring Stars to be inaugurated on Wednesday, designed to highlight, support, and promote inclusive initiatives to encourage equality of participation in outreach and teaching, and at a professional level in the field of astronomy.

Outside of this Convention Center, in different parts of the City of Vienna a large array of different astronomical activities for the general public have been organized: Stargazing Parties; a Public talk at the Austrian Academy of Sciences; Public Talks at the Vienna Planetarium; an International Teacher Training Workshop; Astronomy Youth Day; Gazing into the Cosmos; Exhibition Our Place in Space; Exhibition and Talk on Light Pollution.

The mobile app and the on-line newspaper "Keplersche Nachrichten" or "Kepler's News" will provide us with the latest news, useful information, including the daily schedule and updates needed for business meetings.

On behalf of the International Astronomical Union, let me thank most heartily the National and Local Organizing Committees that have secured for us this beautiful convention center which provides a very convenient environment to carry out our scientific discussions; and also, for the very exciting social activities that have been organized for us to enjoy. I am sure that all the participants will find in this XXX General Assembly many stimulating activities that altogether will make it a memorable experience. Let me wish you a very fruitful scientific meeting.

3. Welcome Remarks

Dr. Piero Benvenuti, IAU General Secretary

Dear Colleagues,
My warmest welcome to the XXXth General Assembly of the International Astronomical Union!

The General Assemblies offer a great opportunity to learn about the most recent and interesting results in all the fields of our beautiful Science, but they are also an occasion to meet old friends, get to know new colleagues – especially the young ones – to exchange ideas and to strengthen or start international collaborations.

If we look back to previous GAs, we can say that each one has been characterised by something special for which it is remembered. Although we should wait for the conclusion of the XXXth GA to know how it will labelled in the history of the IAU, we can anticipate some features that will definitely make the Vienna General Assembly unique.

First of all, the XXXth GA will mark the admission in the Union of more than 300 Junior Members, the new category of individual members – to be approved by the Assembly – that formally allows fresh PhD researchers to actively participate in the life of the IAU. To highlight the special attention that the Executive Committee and the Division Presidents has given to the young generation of astronomers, the GA will salute the 19 winners of the IAU PhD Prize – an award that began in 2016.

Together with the Junior Members, we will welcome more than 1000 new Individual Members, and it is encouraging to note that the gender balance of the new cohort is as high as 30%: a sign that our efforts to promote an egalitarian participation in Astronomy has begun to show its effects. Our community is continuously growing and in the right direction!

I believe the Vienna GA will also be remembered for the exceptionally high number (10!) of new National Members that will join the Union: some of them are re-entering after having been suspended for some years, but the majority are indeed new ones.

An additional element of interest will be the approval of the IAU Strategic Plan 2020–2030: it is not just an update of the existing "Astronomy for Development Strategic Plan 2010–2020," rather it is an attempt to project all the activities of the Union in the next decade. Among the noteworthy new goals of the Plan is the fostering of inclusiveness in the advancement of Astronomy and the proposed creation of the Office for Astronomy Education.

I am confident that all these points, together with great scientific meetings and the foretaste of the celebrations for the 100 years of the IAU (1919–2019) – better described in the welcome address of our President – will make the XXXth a memorable event!

4. Gruber Foundation Presentations

General Secretary Piero Benvenuti introduced representatives from the Gruber Foundation for the Gruber Prize celebrations. The primary goal of the IAU is the development of astronomy worldwide, and so the IAU is pleased to collaborate with the Gruber Foundation on the Cosmology Prize. The Cosmology Prize was established in 2000 as the first Gruber international prize. The Foundation's prize program now presents three annual \$500,000 prizes in the fields of: Cosmology; Genetics; and Neuroscience. Each prize recognizes achievements and discoveries that produce fundamental shifts in human knowledge and culture.

The 2018 Cosmology Prize was awarded to the Planck Team and to Jean-Loup Puget and Nazzareno Mandolesi, the leaders of the High Frequency Instrument and Low Frequency Instrument consortia, for mapping the temperature and polarisation of the cosmic microwave background (CMB) radiation using the ESA Planck spacecraft. The winners were presented their gold medals and gave their thanks for the awards.

The collaboration between the Gruber Foundation and the IAU consists not only of the Cosmology Prize, but also an annual $50,000 Fellowship. The fellowship is administered by the IAU and awarded competitively to a postdoctoral researcher, with the stipend to be used to further his or her research.

The Gruber Fellow postdoctoral awards were announced for the following winners:
2016: Maria Nikolayevna Drozdovskaya and Laetitia Delrez
2017: Khaled Said
2018: Dary Alexandra Ruiz-Rodriguez and Néstor Espinoza

5. PhD Prize Presentations

President Silvia Torres-Peimbert and General Secretary Piero Benvenuti introduced the PhD prizewinners from 2016 and 2017 selected by each Division, and awarded them Certificates. The winners are:

2016:
Division B: Adrian Sven Hamers
Division D: Eric Robert Coughlin
Division E: Shinsuke Takasao
Division F (shared): Laura Kreidberg and Caroline Morley
Division G: Morgan MacLeod
Division H: Long Wang
Division J: Fabio Pacucci
Non-OECD Country: Yang Huang

2017:
Division A : Gisela Ortiz Leon
Division B : Barak Zackay
Division D : Guillaume Voisin
Division E : Christopher Moore
Division F : Megan Ansdell
Division G : Gaël Buldgen
Division H : Georgia Virginia Panopoulou
Division J : Max Gronke
Non-OECD Country : Siyao Xu

The IAU applauds their excellent work.

6. Presentation by the International Science Council

Dr. Heide Hackmann gave a brief presentation on the International Science Council (ISC), created in 2018 as the result of a merger between the International Council for Science (ICSU) and the International Social Science Council (ISSC). The IAU is a member organization of the ISC, which aims to act as the global voice for science and to advance science as aa global public good.

7. Close of Ceremony

Following another musical performance, Gerhard Hensler moved to end the Opening Ceremony and welcomed the General Assembly to the reception that followed.

8. Presentation of Organizers, Supporters, Sponsors and Exhibitors

The IAU and Organising Committees gratefully acknowledge the invaluable support of the following institutions and organisations:

Organisers

International Astronomical Union
University of Vienna
Austrian Academy of Sciences
Austrian Society for Astronomy and Astrophysics

Supporters

Lufthansa
Stadt Wien
Austrian Federal Ministry of Education, Science and Research

Sponsors: Platinum

Vienna Convention Bureau

Sponsors: Gold

Astronomy & Astrophysics
German Astronomical Society
Naturhistorisches Museum Wien

Sponsors: Silver

Korean Astronomy and Space Science Institute
National Academy of Sciences
Chandra X-ray Observatory
National Science Foundation
IOP Publishing
National Astronomical Observatory of Japan
Oxford University Press
Royal Society Publishing

Sponsors: Bronze

The Norwegian Academy of Science and Letters
Nature Astronomy
QYHCCD
Korean Astronomical Society
RadioNet

Sponsors: Friend

Kuffner Sternwarte
Vienna Snowglobe

Exhibitors

DSI - SOFIA
European Solar Telescope
National Radio Astronomy Observatory

International Virtual Observatory Alliance
Korean Astronomical Society
Korean Astronomy and Space Science Institute
Mexico
European Space Agency
MeerKAT
Australian Astronomy
National Science Foundation
Cherenkov Telescope Array Observatory
National Astronomical Observatory of Japan
Long Baseline Observatory
SARAO
European Southern Observatory
Association of Universities for Research in Astronomy
Institute of Space and Astronautical Science & Japan Aerospace Exploration Agency
Square Kilometer Array Telescope
Green Bank Observatory
IPAC at Caltech
Chinese Astronomical Society
Gravitational-Wave Astronomy
International Astronomical Union
Chandra X-Ray Observatory
Southern African Large Telescope
AAO - SCAR AAA
Astronomy & Astrophysics
Cambridge University Press

9. Additional Welcome Speeches

Besides the Welcome Speeches during the Opening Ceremony, there were also greetings in the IAU GA Newspaper from the following dignitaries.

Gerhard Hensler, Chair of the Local Organizing Committee of the IAU XXXth General Assembly

Dear friends and colleagues,

Dear participants at the XXXth General Assembly of the IAU, welcome to Vienna, welcome to the General Assembly of the IAU. These GAs are by far the largest international meetings of astronomers and astrophysicists. Therefore, the Austrian astronomers were delighted and are proud that the Executive Committee of the IAU decided in 2012 at the XXVIIIth GA in Beijing to award the GA 2018 to Vienna.

While the Austrian astronomy has a long tradition beginning at the University of Vienna in the 15th century and highlighted continuously by famous astronomers and by outstanding physicists who contributed substantially to our understanding of the Universe, this GA is special for the contemporary Austrian Astronomers, because it coincides with the 10th anniversary of Austria's accession to ESO. Moreover, this GA is unique to all astronomers as it is closest to the centennial birthday of the IAU next year, so that its pre-celebration includes special events such as the Symposium 349, the centennial ceremony, and an exhibition of astronomical detections and development over this past century.

Although this GA hosts 7 symposia, 15 Focus Meetings and the 9 Division Days, it was impossible to cover the whole spectrum of astronomical topics equally and to fulfill the taste of the entire astronomical community. The IAU Executive Committee and the Division Presidents have done their best to select the most promising of almost 70 meeting proposals. And we, the local organizers, are doing our best to accommodate the meetings and make sure you have the best means to present your scientific research and enjoy the GA. As a special measure to popularize astronomy, for the first time an Inclusion Day is added to the scientific GA. Moreover, several public outreach activities will attract a wide range of people from teacher training to the KinderUni, public talks and night-sky observations.

One cannot foresee all kinds of pitfalls, but we can ensure you that all LOC members and the numerous volunteers are ready to help you solve any problem you might face. We are grateful for your visit and wish you not only fruitful scientific meetings with friends and colleagues in an exciting common atmosphere, but also an enjoyable stay in Vienna, probably with visits to the marvelous landscape of Vienna and Austria and to other attractive cities in Central Europe.

Anton Zeilinger, President of the Austrian Academy of Sciences

Professional astronomy in Vienna shares quite a long common history with the Austrian Academy of Sciences. The building that has served as the Academy's main base since 1857 used to be before the main building of the University of Vienna. On top of this building, the first state-owned astronomical observatory was erected in 1755 at the request of the Jesuit priest Maximilian Hell, who contributed to the measurement of the astronomical unit in 1769.

The Austrian Academy of Sciences has promoted research and outreach in astronomy for many years. It also established a commission for astronomy, and one of its largest institutes, the IWF (lnstitut fur Weltraumforschung) in Graz, is devoted to space research, employing more than 100 researchers from many countries.

As a researcher in the field of quantum mechanics, I have been using telescopes including TNG (Telescopio Nationale Galileo) and WHT (William Herschel Telescope), both on La Palma in the Canary Islands for several experiments on quantum entanglement. I became quite impressed by the technical achievements of astronomical instruments and by the skills of the scientists and engineers designing and operating them.

The General Assembly of the International Astronomical Union to Vienna will witness many novel achievements leading to an understanding of astronomy and astrophysics unimaginable at earlier times. I trust that the meeting will define a wealth of new opportunities which will lead to unprecedented vistas of the Universe in the future.

Simonetta Di Pippo, Director of the United Nations Office for Outer Space Affairs

Welcome to the XXXth General Assembly of the International Astronomical Union! In addition to hosting this year's General Assembly, Vienna is also the home of the United Nations Office for Outer Space Affairs (UNOOSA).

At UNOOSA, we work to bring the benefits of space to humankind by promoting international cooperation in the peaceful uses of outer space and the use of space science and technology for sustainable development.

When it comes to space, we must work together. International cooperation is vital for overcoming barriers and building bridges, and empowering everyone by sharing higher visions, objectives and information. UNOOSA and IAU share a commitment to this and collaborate on a number of initiatives. For example, UNOOSA is helping to make astronomy more accessible to the public through the Open Universe initiative, and IAU has been an active contributor to its preparatory activities. IAU works with UNOOSA in the field of near-Earth Objects and preservation of the "dark and silent" skies, for which we are jointly preparing a conference in 2019. UNOOSA is also the secretariat of the Committee on the Peaceful Uses of Outer Space, with which the International Astronomical Union (IAU) engages as an observer.

In short, IAU is a valued partner of UNOOSA, helping us to carry out our mandate to help everyone, everywhere, access space. I look forward to this partnership growing even stronger in the years to come, and I wish you a very successful and enjoyable General Assembly.

Michael Ludwig, Lord Mayor of Vienna

It is my great pleasure that the Austrian community of astronomers and astrophysicists has succeeded to bring the present XXXth General Assembly of the International Astronomical Union to Vienna, following an initiative of Prof. Dr. Gerhard Hensler. This two-week conference - by far the largest one in this field of research - represents also a highlight for Vienna as an international congress metropolis. The Department of Astrophysics of Vienna University plays a central role in organizing this event. Impressive research results are on the agenda of this General Assembly as well as the discussion of major international projects.

Since about 3,000 participants from 88 countries and from all continents are travelling to Vienna - some of them with their families - our city will gladly present itself as a host city with all its splendid sights, with a wide range of cultural and culinary offers, and will hopefully stimulate further visits to Vienna.

I wish the participants in the XXXth General Assembly of the International Astronomical Union 2018 an exciting and fruitful congress and a nice stay in Vienna.

Transactions IAU, Volume XXX
Proc. XXX IAU General Assembly, August 2018
Debra Elmegreen, ed.
doi:10.1017/S1743921323005604

CHAPTER II

THIRTIETH GENERAL ASSEMBLY

BUSINESS SESSION I

Tuesday, 21 August 2018

Chaired by Silvia Torres-Peimbert, President of the IAU
Austria Center Vienna

1. Opening and Welcome

The President of the IAU, Dr. Silvia Torres-Peimbert, welcomed the delegates and members to this first business session of the General Assembly. The President invited the General Secretary, Dr. Piero Benvenuti, to start the session.

2. Representatives of IAU National Members

The General Secretary listed the representatives of the National Members (IAU member countries). They are given in the table at the end of this Chapter.

The columns list separately:
- the National Members
- the dues category
- the number of votes attached to each National Representative
- the representative for general business matters
- the representative for membership matters

Note that "Interim" National Members or National Members suspended because they are in arrears of their dues do not vote, even if they have appointed a Representative.

The National Representatives for general business matters discuss and vote on financial matters, as well as propose nominations and elect the Finance Committee for the triennium. The National Representatives for membership discuss and vote on Individual Members issues, including nominations and electing the Special Nominating Committee (SNC) and the Membership Committee.

3. Adoption of Agenda

The delegates approved the agenda of the meeting as published in the Program Book.

4. Reminder of Voting Rules

The President reminded the delegates and the members about the voting rules, depending on the nature of the vote, as specified in the Statutes 14 and 15, namely:

Voting at the General Assembly on issues of a primarily scientific nature, as determined by the Executive Committee, is by Individual Members. Voting on all other matters is by National Member. Each National Member authorises a representative to vote on its behalf.

On questions involving the budget of the Union, the number of votes for each National Member is one greater than the number of its category. National Members with Interim status, observer status, or those who have not paid their dues for years preceding that of the General Assembly, may not participate in the voting. On questions concerning the administration of the Union, but not involving its budget, each National Member has one vote, under the same condition of payment of dues as above.

The decisions of the General Assembly are taken by an absolute majority of the votes cast. However, a decision to change the Statutes requires the approval of at least two thirds of all National Members having the right to vote by virtue of article 14.a. Where there is an equal division of votes, the President determines the issue.

To enable the widest possible participation of Individual Members the Executive Committee may decide that voting on certain issues of a primarily scientific nature, as determined by the Executive Committee, shall be open for electronic voting for not more than 31 days counting from the close of the General Assembly at which the issue was raised.

The Executive Committee shall give Members not less than 3 months notice before the opening of the General Assembly of the intention to open certain issues to electronic voting after the General Assembly.

5. Appointment of Official Tellers

The Official Tellers were duly approved:

Nader Haghighipour
Claus Leitherer
Chryssa Kouveliotou

6. Admission of New National Members to the Union

The General Secretary presented the following ten proposed new National Members, who were subsequently unanimously admitted to the Union:

Country: Algeria
Category: I
Adhering Organization: Centre de Recherche en Astronomie Astrophysique et Géophysique

Country: Cyprus
Category: I
Adhering Organization: Cyprus Space Exploration Organisation

Country: Ghana
Category: I
Adhering Organization: Ghana Space Science and Technology Institute

Country: Jordan
Category: I
Adhering Organization: Royal Jordanian Geographic Center

Country: Madagascar
Category: Observer
Adhering Organization: Malagasy Astronomy & Space Science Association

Country: Morocco
Category: Observer
Adhering Organization: Moroccan National Committee for Astronomy - Cadi Ayyad University

Country: Mozambique
Category: Observer
Adhering Organization: Universidade Eduardo Mondlane

Country: Slovenia
Category: I
Adhering Organization: Society of Mathematicians, Physicists and Astronomers of Slovenia

Country: Syria
Category: Observer
Adhering Organization: Syrian Astronomy Association

Country: United Arab Emirates
Category: II
Adhering Organization: Sharjah Center for Astronomy and Space Sciences

7. Report of the Executive Committee (2015–2018)

The report of the Executive Committee was presented by the General Secretary, Dr. Piero Benvenuti, and is presented in Chapter VI of these *Transactions.*

8. Presentation of Revisions to Statutes and Bye-Laws

The revisions of the Statutes to include new categories of Junior Member and Honorary Member, removal of time limits for a National Member to hold Interim or Observer status, and a clause about dissolution of the Union, were presented by the General Secretary and approved. There were no changes to the Bye-Laws. Working Rules are guidelines adopted by the Executive Committee for interpreting the Statutes and Bye-Laws.

The revised Statutes and the current Bye-Laws and Working Rules are presented in Chapter VIII of these *Transactions.*

9. Presentation of the IAU Strategic Plan 2020–2030

Ewine van Dishoeck, Incoming President, and Debra Elmegreen, Vice-President

The IAU Strategic Plan for 2010–2020, adopted at the XXVIIth General Assembly in 2009, was focused on the then-new Office of Astronomy for Development (OAD) using astronomy in developing countries to impact UN Sustainable Development Goals. A resolution at the XXIXth General Assembly in 2015 specified that the Executive Committee should submit for approval at this General Assembly a new Strategic Plan to continue OAD activities. At the Executive Committee Meeting 98 in Mexico City, Mexico, a working group was formed to develop a new strategic plan. In contrast to the 2010–2020 plan, the Strategic Plan for 2020–2030 encompasses all aspects of the IAU, including new activities in the IAU and other IAU Offices in addition to OAD.

The IAU Mission (revised):
The mission of the International Astronomical Union is to promote and safeguard astronomy in all its aspects (including research, communication, education and development) through international cooperation

The process for developing the new strategic plan was as follows:
- Spring 2016: Working group formed by EC 98
  - Debra Elmegreen, Ewine van Dishoeck, Renée Kraan-Korteweg, Piero Benvenuti
- Spring 2017: Solicit input from OAD, OAO, ISYA, DPs, EC
- July 2017: Face-to-face meeting Leiden
- Summer/fall 2017: Writing
- October 2017: Outline sent to entire IAU membership for comments
  - The main suggestion was the addition of an Office of Astronomy for Education (OAE) by Teresa Lago
- Winter 2017/2018: First full draft
- Comments from OAD, EC, DPs
- March 2018: Full draft to National Members
- April 2018: Approval by EC
- May 2018: Post to entire world at IAU website
- May-July 2018: Typeset + print
- August 2018: Proposed for approval at GA (Resolution A1)

The IAU Executive Committee, Divisions, Commissions, and Working Groups focus on the advancement of astronomy. They:
- Stimulate and facilitate sharing of astronomical knowledge among professional astronomers
- Coordinate professional tasks and interactions with other fields
- Recognise excellence in astronomy through prizes
- Diversify portfolio of prizes in next decade

Additional tasks and coordination:
- Astronomical standards
- Naming of astronomical objects and surface features
- Dark and quiet sky protection
- Global coordination of large facilities (just started)

Women in Astronomy
Junior Members (new)
Diversity and Inclusiveness

IAU Offices
The 4 offices (including the proposed OAE) carry out additional aspects of IAU activities. These Offices are:
The IAU Office for Young Astronomers (OYA)
The IAU Office of Astronomy for Development (OAD)
The IAU Office for Astronomy Outreach (OAO)
The proposed IAU Office of Astronomy for Education (OAE)

The IAU Offices each develop detailed Strategic Plans that are consistent with the overall IAU Strategic Plan. The key IAU Office duties are as follows:

The Office for Young Astronomers (OYA) focuses on the training of young astronomers at University level. It organizes the International School for Young Astronomers (ISYA).

The Office of Astronomy for Development (OAD) focuses on the use of astronomy for development by capitalizing on the field's scientific, technological and cultural links and its impacts on society.

The Office for Astronomy Outreach (OAO) focuses on engaging with the public, providing access to astronomical information and astronomy communication. It maintains a network of National Outreach Contacts (NOCs) and amateur groups.

The proposed Office of Astronomy for Education (OAE) will focus on providing training and resources for using astronomy as a stimulus for teaching and education from elementary to high school level (astronomy and science education).

There are mandates for the intersections of OAO, OAE, OYA, OAD with each other and the IAU as a whole, and for interactions with other fields beyond Astronomy.

The new Strategic Goals laid out in the proposed IAU Strategic Plan for 2020–2030 are:

Goal 1: The IAU leads the worldwide coordination of astronomy and the fostering of communication and dissemination of astronomical knowledge among professional astronomers (EC, Divisions, Commissions, Working Groups)

Goal 2: The IAU promotes the inclusive advancement of the field of astronomy in every country (all; OYA; ISYA)

Goal 3: The IAU promotes the use of astronomy as a tool for development in every country (OAD)

Goal 4: The IAU engages the public in astronomy through access to astronomical information and communication of the science of astronomy (OAO)

Goal 5: The IAU stimulates the use of astronomy for teaching and education at school level (the new OAE)

For each goal, a number of actions have been formulated. The OAD, OAO and OAE specific goals and actions will be discussed at GA Focus Meetings and Division Days.

The first action by the IAU Executive Committee is to issue a Call for proposals to form an Office of Astronomy for Education, with these objectives:

Establish a network of National Astronomy Education Coordinators (NAECs)

Analyze use of astronomy in teaching in IAU countries and identify accessible materials and astronomy literacy guidelines (liase with educational ministries and curriculum experts)

Encourage standards for teacher training activities

Organize an annual International School for Astronomy Education (ISAE; for teachers)

Build a database of volunteer IAU members to contribute to teacher training

As we move forward, the IAU depends on its members for successful execution of the Strategic Plan. Be sure to get involved through Working Groups, Commissions, Divisions, and Offices, and encourage colleagues and early career astronomers to join the IAU. Thanks to all those who are already engaged!

10. Presentation of Resolutions

Bruce Elmegreen, Chair of the Resolutions Committee, presented the 5 Resolutions to be voted on at the Business Meeting II. These include one type "A" resolution concerning the administration of the IAU, and 4 type "B" scientific resolutions. The resolutions are presented in their entirety in Chapter V.

For Resolution B4 on a suggested renaming of the Hubble Law, Dr. Elmegreen made the following historical note:

de Sitter (1930) in "On the Magnitudes, Diameters and Distances of the Extragalactic Nebulae, and their Apparent Radial Velocities," (BAN 5, 157) starts by saying:

"Lately several radial velocities of extragalactic nebulae have been published, all of which are large and positive. This makes the question of the distance of these objects of particular importance." [no references given]

. . . 14 pages later, at the end, he states:

"It thus appears that a static solution of Einstein's field equations cannot represent the observed facts. A non-static solution is contained in a paper by Dr. G. Lemaître [reference given], published in 1927, which had failed to attract my notice, but to which my attention was called by Professor Eddington only a few weeks ago. Lemaître finds a solution in which the radius of curvature increases with time. . . [he calculates the constant mass of the universe].

"I hope to return to the discussion of this ingenious solution in a separate communication."

Dr. Elmegreen noted that discussions will take place during the Business Meeting II ahead of the voting.

11. Report of the Special Nominating Committee

The Special Nominating Committee proposed the following slate of nominations for IAU members for the Officers and Members of the IAU Executive Committee for the triennium 2018–2021:

President	Ewine van Dishoeck (NL)
President-Elect	Debra Elmegreen (US)
General Secretary	Maria Teresa Lago (P)
Assistant General Secretary	Ian Robson (UK)
Vice-President	Ajit Kembhavi (IN, 2nd term)
Vice-President	Boris Shustov (RU, 2nd term)
Vice-President	John Hearnshaw (NZ, one term)
Vice-President	Laura Ferrarese (CA, 1st term)
Vice-President	Daniela Lazzaro (BR, 1st term)
Vice-President	Junichi Watanabe (JP, 1st term)
Advisor	Silvia Torres-Peimbert (MX, past President)
Advisor	Piero Benvenuti (I, past General Secretary)

12. Proposals to host XXXII General Assembly, 2024

Four full proposals were received to host the XXXII General Assembly in 2024:

Canada (Montreal)
Italy (Rome)
Mexico (Puebla)
South Africa (Cape Town)

The decision by the Executive Committee was announced at the second business meeting of the General Assembly (see Chapter III).

13. Closure of Session I

With the close of the session, e-voting is open for the NMs to cast their votes for:
Finance Committee
Membership Committee
Special Nominating Committee

There being no other business items to discuss, the President declared the session closed.

COUNTRY:	Category:	Budget Vote Weight:	General Business Matters	Membership Matters	Financial Matters
ARGENTINA	II	3	Cristina Mandrini	Cristina Mandrini	Cristina Mandrini
AUSTRALIA	IV	5	Stuart Wyithe	Stuart Wyithe	Stuart Wyithe
AUSTRIA	II	3	Helmut Rucker	Helmut Rucker	Helmut Rucker
BELGIUM	IV	5	Ronald Van der Linden	Ronald Van der Linden	Ronald Van der Linden
BRAZIL	III	4	Othon Winter	Othon Winter	Othon Winter
BULGARIA	I	2	Lyuba Slavcheva-Mihova	Lyuba Slavcheva-Mihova	Lyuba Slavcheva-Mihova
CANADA	V	6	Kristine Spekkens	Kristine Spekkens	Kristine Spekkens
CHILE	II	3	Patricio Rojo	Patricio Rojo	Patricio Rojo
CHINA Nanjing	VI	7	Xiaowei Liu	Xiaowei Liu	Xiaowei Liu
CHINA Taipei	II	3	You-Hua Chu	You-Hua Chu	You-Hua Chu
COLOMBIA	I	2	Santiago Vargas	Santiago Vargas	Santiago Vargas
CZECH Republic	III	4	Jan Palouš	Jiri Grygar	Miroslav Brož
DENMARK	III	4	Jørgen Christensen-Dalsgaard	Jørgen Christensen-Dalsgaard	Jørgen Christensen-Dalsgaard
ESTONIA	I	2	Laurits Leedjärv	Laurits Leedjärv	Laurits Leedjärv
ETHIOPIA	I	2	Solomon Tessema	Solomon Tessema	Solomon Tessema
FINLAND	II	3	Heikki Salo	Heikki Salo	Heikki Salo
FRANCE	VII	8	Ariane Lançon	Ariane Lançon	Ariane Lançon
GERMANY	VII	8	Joachim Wambsganss	Joachim Wambsganss	Joachim Wambsganss
GREECE	III	4	Athena Coustenis	Athena Coustenis	Athena Coustenis
HONDURAS	INTERIM	OBSERVER			
HUNGARY	II	3	Ágnes Kóspál	Ágnes Kóspál	Ágnes Kóspál
INDIA	VI	7	Dibyendu Nandi	Kanak Saha	Ravi Subrahmanyan
INDONESIA	I	2	Premana Premadi	Premana Premadi	Premana Premadi
IRELAND	I	2	Matthew Redman	Matthew Redman	Matthew Redman
ISRAEL	III	4	Shay Zucker	Shay Zucker	Shay Zucker
ITALY	VII	8	Ginevra Trinchieri	Ginevra Trinchieri	Ginevra Trinchieri
JAPAN	VII	8	Junichi Watanabe	Toshihiro Handa	Hitoshi Yamaoka
Republic of KOREA	III	4	Hyesung Kang	Hyesung Kang	Hyesung Kang
LITHUANIA	I	2	Vytautas Čepas	Vytautas Čepas	Vytautas Čepas
MEXICO	III	4	Susana Lizano	Susana Lizano	Susana Lizano
MONGOLIA	SUSPENDED	OBSERVER			
NETHERLANDS	V	6	Ralph Wijers	Ralph Wijers	Ralph Wijers
NEW ZEALAND	II	3	Nicholas Rattenbury	Nicholas Rattenbury	Nicholas Rattenbury
NORWAY	ii	3	Shahin Jafarzadeh	Shahin Jafarzadeh	Shahin Jafarzadeh
POLAND	IV	5	Andrzej Udalski	Andrzej Udalski	Andrzej Udalski
PORTUGAL	II	3	Rosa Doran	Rosa Doran	Rosa Doran
RUSSIA	V	6	Nikolay N. Samus	Nikolay N. Samus	Nikolay N. Samus
SERBIA	I	2	Bojan Arbutina	Bojan Arbutina	Bojan Arbutina
SOUTH AFRICA	III	4	Petri Väisänen	Petri Väisänen	Petri Väisänen
SPAIN	IV	5	Rafael Bachiller	Rafael Bachiller	Rafael Bachiller
SWEDEN	III	4	Dainis Dravins	Dainis Dravins	Dainis Dravins
TAJIKISTAN	I	2	Kokhirova Gulchehra	Kokhirova Gulchehra	Kokhirova Gulchehra
THAILAND	I	2	Busaba Kramer	Busaba Kramer	Busaba Kramer
TURKEY	I	2	Emre Işık	Kadri Yakut	Kadri Yakut
UKRAINE	III	4	Akhmetov Vladimir	Akhmetov Vladimir	Akhmetov Vladimir
UNITED KINGDOM	VII	8	Mike Cruise	Mike Cruise	Mike Cruise
UNITED STATES	X	11	Kenneth Kellermann	Kenneth Kellermann	Kenneth Kellermann
VATICAN	I	2	Christopher Corbally	Richard Boyle	Pavel Gábor
CYPRUS	New NM	OBSERVER			
JORDAN	New NM	OBSERVER			
MOROCCO	New NM	OBSERVER			
SLOVENIA	New NM	OBSERVER			
UAE	New NM	OBSERVER			

Figure 1. National Members and their Representatives (see §2). Credit: Piero Benvenuti, with the assistance of Rosaria d'Antonio and Madeleine Smith-Spanier

Transactions IAU, Volume XXX
Proc. XXX IAU General Assembly, August 2018
Debra Elmegreen, ed.
doi:10.1017/S1743921323005616

CHAPTER III

THIRTIETH GENERAL ASSEMBLY

BUSINESS SESSION II

Thursday, 30 August 2018

Chaired by Silvia Torres-Peimbert, President of the IAU
Austria Center Vienna

1. Opening and Welcome

The President of the IAU, Dr. Silvia Torres-Peimbert, welcomed the delegates and members to the second business session of the General Assembly. The President invited the General Secretary, Dr. Piero Benvenuti, to start the business session. The agenda was approved.

2. Individual and Junior Membership

The General Secretary reported that, after screening by the Membership Committee, 1240 new members were accepted, including 888 Individual Members and 352 Junior Members. The list of new Members was projected on the screen. Chapter VIII presents the Report of the Membership Committee, including the lists of the names of Individual and Junior Members.

3. Honorary Members

For the first time, the IAU has Honorary Members. These are individuals that have substantially contributed to the development of astronomical research and culture in their country, who are not professional astronomers. They are nominated by National Members and approved by the Executive Committee. The first 9 Honorary Members proudly admitted at this General Assembly are:

Czech Republic: Jiří Dušek, Director, Brno Planetarium

Egypt: Hatem Hamdy Odah, President, National Research Institute of Astronomy and Geophysics

Ethiopia: Tefera Waluwa Wondimegegn, Deputy Prime Minister (2003–2010), Founder of Ethiopian Science Society

France: Christian Buil, Optical Senior Expert, CNES

Hungary: Attila Mizser, General Secretary, Hungarian Astronomical Society

Ireland: Joe Hogan, Open Telecom, Founder and CTO

Russia: Olga Y. Vasilyeva, Minister of Education and Science (2016)

Ukraine: Gordienko Sergey Pavlovich, Founder and Editor-in-chief of "The Universe, Space, and Time"

United States: Wayne Rosing, Chief engineer, Director, Las Cumbres Observatory

Their citations are available on https://www.iau.org/administration/membership/honorary/.

4. Deceased Members

The Executive Committee regretfully reported 348 deceased Individual Members of the Union, since August 2015. The list includes three former IAU Presidents, Yoshihide Kozai (1988–1991), Alexander A. Boyarchik (1991–1994), and Robert P. Kraft (1997–2000). The General Secretary displayed all of the names on slides, and the General Assembly observed a one-minute silence in their respectful memory. The complete list of deceased members is given in Chapter IX.

5. Appointment of the Official Tellers

The Official Tellers were duly appointed:

Rosa Duran (P)
Claus Leitherer (USA)
Denis Burgarella (F),

who confirmed that the quorum requirements for the National Representatives were satisfied.

6. Changes to Statutes and Bye-Laws

The revisions to the Statutes, as reported at the Business Meeting I, were approved; namely, to add new membership categories of Junior Member and Honorary Member, to remove time limits for Interim and Prospective (now changed to "Observer") status, and to give detailed prescriptions in case of dissolution of the Union (necessary for obtaining the Charity Equivalency Status).

7. Resolutions

The Resolutions submitted to the General Assembly were examined by the Resolutions Committee, chaired by Bruce Elmegreen. One was "A" type, having to do with the administration of the IAU. Four were "B" type, i.e., with scientific content. They were submitted in English and published in the GA Newspaper before the vote. The French version was made available on the IAU web site after their adoption. The Resolutions are presented in their entirety in Chapter V.

As reported at Business Meeting I, these Resolutions were:

A1 on the IAU Strategic Plan 2020–2030
B1 on Geocentric and International Terrestrial Reference Systems and Frames
B2 on The Third Realization of the International Celestial Reference Frame
B3 on preservation, digitization and scientific exploration of historical astronomical data
B4 on a suggested renaming of the Hubble Law.

After discussion and clarifications, the first four Resolutions A1, B1, B2, and B3 were unanimously approved. Resolution B4 generated much discussion. Based on concerns expressed by members ahead of the General Assembly, the Executive Committee decided in advance that a straw poll would be taken at the General Assembly, (which indicated approval), but that there would be an electronic vote by the full IAU membership. Votes were accepted until Oct. 26, and the resolution was ultimately approved. In Chapter V,

Resolutions Committee Chair Bruce Elmegreen details the discussion, questions, and answers leading up to the final vote.

8. Vote on proposal to modify Division F name

The Steering Committee of Division F, "Planetary Systems and Bioastronomy," is proposing to modify the name of its Division to "Planetary Systems and Astrobiology." Although the two terms "Bioastronomy" and "Astrobiology" can be considered synonyms, only the latter can be found in the most authoritative English Dictionaries as the most commonly used and correct term to indicate the field of science encompassing Astronomy and Biology concerned with the origin, early evolution, distribution and future of life in the universe. The Executive Committee, in its Meeting #99 (Pune, 2017), has endorsed the proposal that is now presented to the General Assembly for formal approval.

The proposed name change was approved by the Congress.

9. Elected Division Presidents and Vice-Presidents

The General Secretary presented the names of the Division Presidents and Vice-Presidents that were elected by the Members of their respective Divisions. The percentage of Division members voting ranged from 19% to 25%.

The new leaders are as follows:

Division A: Fundamental Astronomy
President: Daniel Hestroffer (France)
Vice-President: Norbert Zacharias (USA)

Division B: Facilities, Technologies, and Data Science
President: Michael Burton (Australia)
Vice-President: Gabriele Giovanni (Italy)

Division C: Education, Outreach, and Heritage
President: Susana Deustua (USA)
Vice-President: Richard de Grijs (Australia)

Division D: High Energy Phenomena and Fundamental Physics
President: Elena Pian (Italy)
Vice-President: Isabelle Grenier (France)

Division E: Sun and Heliosphere
President: Sarah Gibson (USA)
Vice-President: Cristina Mandrini (Argentina)

Division F: Planetary Systems and Bioastronomy
President: Gonzalo Tancredi (Uruguay)
Vice-President: Maria Antonella Barucci (France)

Division G: Stars and Stellar Physics
President: David Soderblom (USA)
Vice-President: Andrej Prsa (USA)

Division H: Interstellar Matter and Local Universe
President: Leonardo Testi (Italy)
Vice-President: Monica Rubio (Chile)

Division J: Galaxies and Cosmology
President: : Denis Burgarella (USA)
Vice-President: Kim-Vy Tran (Australia)

The Assembly took note.

10. Elected Commission Presidents

The General Secretary presented the names of the Commission Presidents who were elected by the Members of the respective Divisions.

A1: Jean Souchay (France)
A2: Florian Seitz (Germany)
A3: James Lindsay Hilton (United States)
A4: Alessandra Celletti (Italy)
X2: Giovanni B. Valsecchi (Italy)

B1: Dmitrij V. Bisikalo (Russian Federation)
B2: Anja C. Schröder (South Africa)
B3: Prajval Shastri (India)
B4: Anthony James Beasley (United States)
B5: Helen J. Fraser (United Kingdom)
B6: Antonio Mario Magalhaes (Brazil)
B7: Constance Elaine Walker (United States)

C1: Paulo Sergio Bretones (Brazil)
C2: Richard Tresch Fienberg (United States)
C3: Wayne Orchiston (Australia)
C4: Gudrun B. E. Wolfschmidt (Germany)

D1: Marica Branchesi (Italy)
X1: Christine Jones (United States)

E1: Alexander Kosovichev (United States)
E2: Paul Cally (Australia)
E3: Nicole Vilmer (France)

F1: Diego Janches (United States)
F2: Jack Lissauer (United States)
F3: Masatoshi Ohishi (Japan)
X2: Giovanni B. Valsecchi (Italy)

G1: Virginia Trimble (United States)
G2: Jorick S. Vink (United Kingdom)
G3: Marc Pinsonneault (United States)
G4: Jaymie Matthews (Canada)
G5: Carlos Allende Prieto (Spain)

H1: Dante Minniti (Chile)
H2: Edwin A. Bergin (United States)
H3: Albert Zijlstra (United Kingdom)
H4: Amanda Karakas (Australia)

J1: Cristina Carmen Popescu (United Kingdom)
J2: Valentina D'Odorico (Italy)
X1: Christine Jones (United States)

11. Financial Matters

Report of the Finance Committee, presented by João Alves (Chair)

The Finance Committee received the IAU Financial Projection for 2019–2021 on the 13th February and had a video conference a month later on the Projection. The Financial Report 2017, together with the Accounting Report 2017, the Audit Report, and the Special Audit Report reached the Committee on the 11th April. The Committee members have discussed the report via emails. They conclude that:

1) The projection for the next triennium is reasonable, and the Committee supports it.
2) The Financial Report presents details of financial transactions, a summary of the external auditing report, analyses of assets, incomes, and expenditures.

Details of the budget were given by General Secretary Piero Benvenuti, and tables are presented Chapter VI in this Volume).

The report was submitted to the vote of National Representatives, and the triennium budget proposal for 2018–2021 was approved.

12. Election of the Finance Committee

The General Assembly appointed the following Finance Committee (2018–2021):

João Alves (Austria), Chair
Kate J. Brooks (Australia)
Michel Dennefield (France)
J.J. Matthew Kavelaars (Canada)
László L. Kiss (Hungary)
Na Wang (China)
Tushar P. Prabhu (India)
Lee Anne M. Willson (USA)

13. Election of the Membership Committee

The General Assembly unanimously appointed the following Membership Committee (2018–2021):

David Soderblom (USA), Chair
Soňa Ehlerová (Czech Republic)
Laurent Eyer (Switzerland
Sofia Feltzing (Sweden)
William E. Harris (Canada)
Dibyendu Nandi (India)
Serena Viti (Netherlands)

14. Appointment of the Resolutions Committee

The General Assembly unanimously appointed the following Resolutions Committee (2018–2021):

Sergei A. Klioner (Germany), Chair
Katia Cunha (Brazil)
Toshio Fukushima (Japan)
Claus Leitherer (USA)
Boris M. Shustov (Russia)

15. Appointment of the Special Nominating Committee

The General Assembly unanimously appointed the following Special Nominating Committee (2018–2021):

Members:
Ewine van Dishoeck (President, Netherlands; Chair)
Jonathan Bland-Hawthorn (Australia)
Beatriz Elena García (Argentina)
Masahiko Hayashi (Japan)
Gražina Tautvaisiene (Lithuania)
Silvia Torres-Peimbert (Past President, Mexico)
Lidia van Driel-Gesztelyi (UK)

IAU Advisors:
Teresa Lago (General Secretary; Portugal)
Ian Robson (Assistant General Secretary; UK)

16. Election of the Executive Committee 2018–2021

The General Assembly unanimously appointed the following Executive Committee (2018–2021):

President	Ewine van Dishoeck (Netherlands)
President-Elect	Debra Elmegreen (USA)
General Secretary	Teresa Lago (Portugal)
Assistant General Secretary	Ian Robson (United Kingdom)
Vice-President	Laura Ferrarese (Canada)
Vice-President	John Hearnshaw (New Zealand)
Vice-President	Ajit Kembhavi (India)
Vice-President	Daniela Lazzaro (Brasil)
Vice-President	Boris Shustov (Russia)
Vice-President	Junichi Watanabe (Japan)
Advisor	Silvia Torres-Peimbert (Mexico)
Advisor	Piero Benvenuti (Italy)

17. Special Executive Committee Award for Astronomy Outreach, Development, and Education

The IAU Strategic Plan (2020–2030) calls for the IAU to continue to enhance and diversify its portfolio of prizes to reflect changes in its priorities. The Executive Committee aims to have regular prizes in outreach, development and education. This special award is a first step, generously provided by incoming President Ewine van Dishoeck.

Ewine van Dishoeck:

The 2018 special prize is awarded on behalf of the IAU Executive Committee to Dr. Carolina Ödman, Associate Professor at the University at the Western Cape and Associate Director, Development & Outreach for the Inter-University Institute for Data Intensive Astronomy in South Africa, "for her pioneering work in astronomy outreach, development and education."

During the last 15 years Carolina has used her unique combination of creativity, intelligence, pragmatism, and warm personal qualities to develop programmes that use astronomy to benefit society and humanity globally. Her accomplishments are many and unique.

She received an Astrophysics PhD in theoretical cosmology from Cambridge UK in 2003. Shortly thereafter, she joined Leiden University in 2005. Her task was to explore the possibility of initiating a new educational outreach programme, to inspire very young economically disadvantaged children with the beauty and size of the Universe. The idea of "Universe Awareness" was to use astronomy to help stimulate tolerance at a formative stage of a child's development. As UNAWE International Project Manager from 2005 until 2010, Carolina turned Universe Awareness from an idea into a thriving international programme with a multidisciplinary network of hundreds of astronomers, teachers and outreach experts, operating in more than 40 countries, reaching more than 200,000 children in that period. It was a global cornerstone of the International Year of Astronomy. The success of UNAWE in these early years is in large part due to Dr. Ödman's exceptional talents as well as her perseverance and hard work.

Shortly afterwards, she emigrated to South Africa, a country that has pioneered the use of astronomy as a tool for capacity building. She became Director of Academic Development at the AIMS Next Einstein Initiative at the African Institute for Mathematical Sciences in Cape Town. There she contributed substantially to the AIMS mission of giving bright young African scientists an opportunity to learn from some of the best lecturers around the world.

In 2013 Dr. Ödman became Chief Scientist at Thumzup, an innovative South African company that pioneers the development of new technology for consumers. Recently she re-joined the astronomical community as Associate Professor at the University of the Western Cape and Associate Director, Development & Outreach at the Inter-University Institute for Data Intensive Astronomy - a partnership of four South African universities.

It is clear that Carolina has been a pioneer in all aspects of education, outreach and development, and that she is a highly worthy recipient of this prize.

Unfortunately Carolina Ödman cannot be in Vienna to receive the prize in person, but it can be presented to her husband, Kevin Govender, the Director of the IAU Office of Astronomy for Development, to which she has also contributed her advice and energy.

18. Date and location of the XXXII General Assembly in 2024

It was announced that the XXXII General Assembly will take place in Cape Town, South Africa, in August, 2024. This will be the first time for a General Assembly to be held in Africa.

Transactions IAU, Volume XXX
Proc. XXX IAU General Assembly, August 2018
Debra Elmegreen, ed.
doi:10.1017/S1743921323005628

CHAPTER IV

THIRTIETH GENERAL ASSEMBLY
CLOSING AND HANDOVER CEREMONY

Thursday, 30 August 2018

Austria Center Vienna

1. Dance I

The Closing and Handover Ceremony began with a group photo, followed by a traditional folk dance and a welcome from the Master of Ceremony and incoming President Ewine van Dishoeck.

2. Closing Address by the Retiring IAU President

Dr. Silvia Torres-Peimbert

Dear friends, It has been a great honor for me to serve the Union as President for the past three years. Like any organization that aspires to be useful to its constituency it requires continued supervision. I am very grateful to those that have participated in the administration and development of the Union. The General Secretary carried a big responsibility in running this organization. In this task the President Elect and the Assistant General Secretary have participated continuously. The members of the Executive Committee, the Division Presidents, Commission Presidents and Working Group Presidents, as well and the many Steering Committees have contributed in different tasks. All these colleagues generously have strengthened this organization with their time and ideas. In addition, I want to recognize the Secretariat that has worked very efficiently to accommodate all the ever-increasing demands from all of us.

I am personally delighted by the selection of the next Officers and Executive Committee that under the direction of Ewine van Dishoeck will lead this organization to higher levels.

Again, I wish to express my warmest thanks and deep appreciation to our Austrian colleagues, the National and the Local Organizing Committee, under the leadership of Prof. Gerhard Hensler, for the impeccable running of the General Assembly and its convivial atmosphere.

Next year is the centennial anniversary of the IAU, and I expect that all of us upon returning to our home town will spread the enthusiasm to celebrate this event. I wish you all a safe trip home and I expect to see you all in 2021 in Busan for the XXXI IAU General Assembly.

3. Closing Address by the Retiring General Secretary

Dr. Piero Benvenuti

Dear Colleagues, dear Friends, there is a substantial difference between the incoming talk of a General Secretary and his outgoing one: the first may be interesting, it talks about the future, it opens up expectations and occasionally can even provoke viral criticism on social networks...

On the contrary the latter risks to be extremely boring, particularly since it comes after two weeks during which I have been pestering you with Statutes, Rules, crazy Resolutions...

What has been done during the past triennium is in front of you, you can judge by yourself if it was good or bad without much else to add – the only thing I want to highlight is that I enjoyed very much working with my colleague Officers, Silvia, Ewine and Teresa: personally and professionally it has been a remarkable high time.

Anyway, it is not my intention to waste the unique opportunity I have to talk to such a large audience of distinguished colleagues, therefore I decided to turn my outgoing speech into an incoming one.

Incoming in my new, craved position of retired astronomer. Retired from my University, three years ago, retired from the position of General Secretary, few minutes ago.

I will tell you about my future personal plans, which however are well related to the fantastic experience of the past three years, during which I had the opportunity to encounter so many diverse cultures, all united by the same passion for our beautiful science, astronomy.

I rediscovered, if ever proof were needed, that astronomy knows no barrier of any sort, because it is deeply rooted in our aspiration in a better world and in our significant position in the cosmos, beyond the space-time constraints.

But then, coming back to my western culture, I suddenly realized how much our cosmological knowledge has lost the profound impact it had in the ancient times. Today, philosophy and theology continue to proceed in their quests as if the Copernican revolution never happened.

We know how this occurred: when Galileo, in the fatal nights of the winter of 1609 opened the new window on the sky, he also shattered down the crystalline spheres of the Aristotelian cosmology, but obviously he could not replace it with an equally satisfactorily model of the world. Only today, I would say after the longed detection of the gravitational waves and the fantastic ALMA images of the planetary systems' formation, we can say that the Copernican revolution is complete.

But philosophers and theologians seem to ignore the fact and keep reasoning in terms of absolute concepts of space and time, giving more relevance to the "being" rather than to the "relationship," while we urgently need a revised philosophy of nature that develops a solid ontology of our "entangled" world, in which the classic distinction among rocks, plants, animals and conscious beings is replaced by a unified network of relations.

We, the scientists, are partly co-responsible for the current unsatisfactorily situation, possibly because we are insisting too much, in our outreach actions, in the purely scientific content of our discoveries, with less attention to its far reaching global cultural consequences.

This is the challenge I wish to pursue in my new job, rocking the boat of world religions, in particular of Christian theology, following the path traced by my masters: Andrea Zanzotto, a great Italian poet and my school teacher, George Lemaître, who saved a Pope from saying naiveties about Creation and Pierre Teilhard de Chardin.

I may even propose an IAU Symposium on the subject for Busan, so that eventually I will enjoy, as I hope you all did here in Vienna, the real participation in a General Assembly!

Therefore: Auf-widersehen in Busan!

I cannot leave the podium and the General Secretary position, without expressing a couple of much deserved acknowledgements. The first to Rosaria and Madeleine for their professional and personal dedication to the Union and for the understanding and support they gave to my crazy mini-revolution in the management of the Office. They are now not just IAU "employees," they are the backbone of the Union... and, just for the record, the video about the IAU that you watched yesterday is essentially the work of Rosaria.

However, the second greatest thank goes to my wife Beatrice; during these three years she sustained me with her wisdom and sometimes with her very justified and constructive criticism. Without her constant help I would not have been able to complete my task maintaining mental sanity... And during the last three months, when I became essentially invisible, both physically and psychically, she refrained from asking divorce just because she loves me dearly: thank you!

4. Address by the Incoming President

Prof. Ewine van Dishoeck

Dear colleagues and friends,

It is a great honor for me to address you as the new President of the IAU. Together with my husband Tim de Zeeuw, I attended my first IAU General Assembly in 1979 in Montreal at the invitation of the then President, Adriaan Blaauw. I was still a MSc student in theoretical chemistry, but it was a fantastic experience to be there, seeing and talking to all these famous astronomers like Chandrasekhar! I could never have dreamed that I would be standing here in front of you 40 years later.

The IAU organization depends on astronomers dedicating their "free" time to make the IAU work, and over the last three years we have been fortunate to have an extremely capable group at the top: Silvia Torres-Peimbert as the "heart" of the

IAU, Piero Benvenuti working night and day as the GS supported in the Paris Office by Rosaria d'Antonio and Madeleine Smith-Spanier, Teresa Lago having an increasing load as AGS, and with overall help from very active Vice Presidents and Division Presidents. Also, the extremely capable support by the IAU press officer Lars Lindberg Christensen and the ESO team on website and databases is essential. Many thanks to all of you; the biggest applause should undoubtedly go to Piero. I look forward to working closely with the new Officers, Teresa, Debra Elmegreen and Ian Robson, and the new Executive Committee and Division Presidents, to hopefully make the running equally smooth. With three, and soon four, IAU Offices, this is no small enterprise compared with a decade ago. In the Strategic Plan we quoted the Indian philosopher Rabindranath Tagore: "You can't cross the sea by simply standing and staring at the water." We need the help and support of many of you to make our dreams into reality.

So, "what are you going to do as President?," I got asked a lot in the corridors. When I became President-elect three years ago I had a number of goals. The first of them was to promote the presence of young scientists in the Union, and I was pleased to find at our first meeting in Paris that the other Officers had exactly the same priority. So now we have the Junior Member category. The challenge will be for the Junior Member WG to develop activities worldwide that benefit their development and to learn best practices from each other. They can also help to promote astronomy worldwide, and I look forward to working with them. A second goal was to provide a more comprehensive overview of the IAU as a whole: where do all our activities, many of them new, fit in the bigger picture? What are we missing and where are overlaps? What are the mandates of the various Offices? This has resulted in the writing of the Strategic Plan 2020–2030, mostly together with Debra, so it will be our joint responsibility that the list of actions in that plan are implemented. As an example, further promoting Women in Astronomy, and more generally Diversity and Inclusion in the IAU, is high on that list.

Celebrating the IAU 100 years will keep me very busy in 2019, but it is also a fabulous opportunity to give many of our Strategic Plan goals a boost and show to our fellow scientists, policy makers, teachers, educators and general public what a century of astronomy has brought to society. Fundraising both for IAU100, OAD and IAU as a whole will be part of this.

Another area that I feel strongly about is the Global Coordination of Large Facilities, a WG that has been elevated to report directly to the Executive Committee. Astronomy is an observationally driven field, so progress requires new increasingly larger telescopes and instrumentation that can see sharper, deeper and more sensitive than previous facilities. Astronomy has been fortunate to have an armada of new telescopes and satellites across the electromagnetic spectrum over the past decades, with multimessenger facilities ramping up. But this steady stream of new facilities cannot be taken for granted. We are already seeing that some of our dreams, like the 20–40m class telescopes on the ground, are taking longer to come to fruition. Large scientific infrastructures are no longer the domain of just the astronomers and the physicists: other areas such as the Atmospheric Science, Geochemistry, Life sciences, and even the Humanities and Social Sciences, require increasingly large facilities as well (often in the form of large databases) to advance their field. They now also have the ears and eyes of the funding agencies and politicians (don't underestimate this!). Global collaboration or coordination in astronomy is therefore increasingly important as the costs and preparation to build, and especially operate, our forefront facilities and missions escalate. The IAU provides a neutral forum to bring various parties together and foster collaboration at an early stage. With support of the Kavli foundation we will be stepping up our WG activities.

This brings me to the final aspect that I would like to bring to your attention. The wheel that was on the cover of the 2010–2020 Strategic Plan, and that is also included in the 2020–2030 plan, illustrates in a powerful way our links with other fields, ranging from technology and large data, to other sciences and culture. Without often realizing it, astronomy is very dependent on developments and work in these fields. Without fundamental physics and material sciences we would not have the progress in detectors, telescope components and control, etc. A program like CLOUDY, used by hundreds of astronomers, required a century of work and funding by thousands of atomic and molecular physicists to get all the input basic data: spectroscopy, oscillator strengths, collisional rate coefficients. More generally, science is becoming more and more inter- or multi-disciplinary, and a shift is happening already at funding agency level, with new funding going primarily in that direction. As a field, we need to be prepared for this shift, and this means fostering and enhancing collaborations and making scientists from other fields feel welcome in our midst. This is another form of Inclusion, but equally important. I would therefore like to encourage more interdisciplinary IAU Symposia in the coming years, on average at least one per year. We have seen an excellent example here at the GA with the Origins symposium, with speakers from geochemistry, atmospheric science and biology. The 2016 Gravitational waves symposium was another example, together with physics, and in 2019 there will be the first IAU Laboratory Astrophysics symposium. Perhaps a next one could be on big data together with informatics.

Finally, I would like to end with some wise words from the past. Antoine de Saint Exupéry wrote: "If you want to build a ship, don't drum up people to collect wood and don't assign them tasks and work, but rather teach them to long for the endless immensity of the sea." Taking the freedom to modify this quote to apply to the IAU, it becomes: "If you want to do astronomy, don't drum up people to collect components and don't assign them tasks and work, but rather teach them to long for the endless immensity and beauty of the Universe." Ultimately it is this inspiration, this quest to determine our place in the Universe, what it is all about and that drives us all: scientists, engineers and technicians alike. I look forward to take the IAU into its second century with you in the coming years.

5. Address by the Incoming General Secretary

Prof. Teresa Lago

Dear colleagues, I feel privileged to become IAU General Secretary today. It is an exciting time to get closer involved with IAU. Let me explain why: In terms of science, IAU is going through a particularly healthy period. My reasons for this statement are based on one aspect of my recent experience as AGS - the duties associated with the IAU meetings. The last 3 years have been unique, even record breaking, in terms of competition to organise scientific meetings under the umbrella of the IAU:

- 33 proposals for IAU Symposia for 2017, although only 9 could be supported,
- 69 proposals for Symposia & Focus Meetings for 2018, of which we could only select 3 Symposia outside the GA, and 7 Symposia plus 15 FM for the GA programme,
- 30 proposals for IAU Symposia for 2019, from which 9 could be selected.

I see this as a signature of a vibrant community.

This sentiment was further strengthened by the competition to organise the 2024 General Assembly: IAU received 7 Letters of Intent. Once again, a record breaking which led to the need to adopt a two-step selection process in order to avoid wasting time and money on the part of the proponents and also to allow for the proper discussion and evaluation of the full proposals. The more complete information requested from bidders was used for a pre-selection and subsequently the complete proposals were requested only from the four pre-selected bidders.

This large number of proposals reflects the relevance of IAU to member countries.

There is yet another reason for my enthusiasm: the IAU Strategic Plan for 2020–2030. This is the first time IAU has prepared and approved a Strategic Plan that clearly defines the mission - to promote and safeguard astronomy in all its aspects through international cooperation - and goes further giving a comprehensive overview of the Union's general activities, how they fit together and how they complement each other. The goals proposed for the decade are:
- global coordination of Astronomy,
- the dissemination of knowledge among professional astronomers and the inclusive advancement of Astronomy in every country,
- the use of Astronomy as a tool for development and for education,
- the engagement of the public with Astronomy through communication.

Allow me to emphasise the key words: coordination, dissemination of knowledge, inclusive advancement, tool for development and for education, engagement of the public. I take these keywords as the guidelines for my mandate as General Secretary of IAU.

I foresee the next three years not just in my role as General Secretary, but as a shared mission involving each and every one of you as well. I count on your initiative, your passion and your commitment to Astronomy.

To fulfill the ambition expressed in the Strategic Plan I perceive IAU as a solid organisation that encompasses not only all its scientific structures - Divisions, Commissions and Working Groups - but also consolidates on its pillars: the four offices dedicated to the pursuit of Education through Astronomy, the training of future astronomers through ISYA, Astronomy for Development and the Communication of Astronomy to the general public. Three of these Offices have been established through bilateral agreements and are already in operation. To install the fourth office - Astronomy for Education - an international call will soon be launched to identify the IAU partners in this new adventure. The goal is that by the middle of the triennium, the four offices will be fully operational, involved in collaboration with each other, and sharing a similar standing within the organisation.

IAU is approaching 100 years. The organisation grows older pursuing its long-term traditions. But it also grows in size with new members, grows ambitious with bold proposals and rejuvenates by opening up to the younger generation of astronomers.

I have learned from experience that dreams can come true through determination, creativity, and above all, networking. I feel that this endeavour will have to be global, involving members and structures, together to fulfil the IAU mission. Together, we will advance and excel, bounded by clear principles: transparency, inclusiveness - the many types of inclusiveness - and creativity.

As you see this is really an exciting time! And my ambition is, over the next three years, to use Astronomy and IAU initiatives as an enlightening and binding tool for realisation, all over the world. Together under one sky! I will do my best. And I count on you!

6. Dance II

A traditional dance was performed.

7. Closing Remarks

Gerhard Hensler, Chair of the NOC

Closing remarks were made by the NOC and LOC, with thanks to the organizers and volunteer helpers, who were invited onstage.

8. Invitation to the XXXIst General Assembly 2021

Hyesung Kang, Chair of the National Organizing Committee for the XXXIst General Assembly

The Korean Astronomical Society would like to welcome you to the IAU XXXI General Assembly, which will be held in Busan, Republic of Korea, in 2021. During the bidding competition in 2015, we set the main theme of the Busan GA as "Astronomy for All," in a sincere hope to make this event a true festival for astronomers from all around the world. To that end, the KAS pledged a special travel grant, which will complement the traditional IAU travel grant, to encourage more young students and junior postdocs to attend the Busan GA.

Let me take a moment to introduce the host city, Busan, situated on the southeastern coast of the Korean peninsula. It has earned recognition as the best place to visit in Asia in 2018 by Lonely Planet.

You will find beautiful beaches, gentle hills, Buddhist temples, and abundant tourist attractions where contemporary lifestyle meets long-standing history. Though summer days can be warm, you can relish the cool sea breeze at Haeundae Beach.

Our venue, BEXCO, stands out as the most competitive exhibition and convention center in Northeast Asia, boasting a state-of-the-art auditorium, ample meeting rooms, and exhibition space. The area also provides plenty of hotels at affordable prices, convenient shopping centers, and an advanced IT infrastructure, ensuring a comfortable and productive experience for all attendees.

To give you a glimpse of Busan's rich natural, cultural, and urban heritage, we have prepared two short videos. Notably, the first movie was produced by college students who enthusiastically wished to showcase the beauty of their proud country.

Let us carry this momentum forward and look forward to the "Astronomy for All" festival at the 31st General Assembly in Busan in 2021.

9. IAU Flag Handover Ceremony

Two videos welcomed participants to the XXXIst General Assembly to be held in Busan, Republic of Korea. Gerhard Hensler, Taubner and Wallner passed the IAU flag to Hyesung Kang as incoming Chair of the NOC for Busan, and Ewine van Dishoeck as incoming IAU President.

10. Dance III

A traditional waltz was performed, and guests were invited to join in onstage and in the audience as the Congress closed in dance.

11. Group Photo and Image Archive

Images taken during the General Assembly can be viewed and dowloaded from the IAU web site:
`https://www.iau.org/public/images/archive/category/general_assembly_2018/`

Figure 1. Participants at the XXXth General Assembly, Austria Center, Vienna. Credit: IAU/M. Zamani

Transactions IAU, Volume XXX
Proc. XXX IAU General Assembly, August 2018
Debra Elmegreen, ed.
doi:10.1017/S174392132300563X

CHAPTER V

THIRTIETH GENERAL ASSEMBLY

RESOLUTIONS OF THE XXXth GENERAL ASSEMBLY

1. Resolutions Committee (2015–2018)

The members of the Resolutions Committee for the 2015–2018 triennium were:

Bruce Elmegreen (USA; Chair)
Renée Kraan-Korteweg (South Africa)
Katia Cunha (Brazil)
Toshio Fukushima (Japan)
Sergei Klioner (Germany)

2. Approved Resolutions

RESOLUTION A1

on the IAU Strategic Plan 2020–2030

Proposed by the IAU Executive Committee

The XXX General Assembly of the International Astronomical Union,

considering

1. That the XXIXth General Assembly, meeting in Honolulu, Hawai'i, USA, August 2015, resolved that the "Strategic Plan: Astronomy for the Developing World" should continue until the XXXIst General Assembly to be held in August 2021 in Busan, Republic of Korea.

2. That the Executive Committee should present for approval at the XXXth General Assembly to be held in Vienna, Austria in August 2018 an extended Strategic Plan which addresses the future of the OAD and its activities beyond 2021.

3. That the Executive Committee should consult existing and potential stakeholders in the preparation of this Strategic Plan.

4. That the Executive Committee, in its 98th annual meeting in Mexico City in May 2016 appointed a dedicated Working Group for the preparation of the extended Strategic Plan.

5. That the Executive Committee, in the same 98th meeting, decided that the extended Strategic Plan, called the "IAU Strategic Plan 2020–2030", should encompass, beyond the Astronomy for Development aspects, all the activities of the IAU.

6. That in the course of 2017 the Working Group solicited inputs from the OAD, OAO and OYA/ISYA, along with the EC Members and Division Presidents and drafted a preliminary version of the IAU Strategic Plan 2020–2030.

7. That in October 2017 the Working Group, before submitting the draft Strategic Plan to the National Members, sounded the opinion of the IAU community at large about the completeness of the Strategic Plan and collected and analyzed more than 150 comments by IAU Individual Members.

8. That in February 2018 the revised draft was distributed for comments to all the IAU National Members representatives.

9. That the final draft of the Strategic Plan 2020–2030 was approved by the Executive Committee in its 100th annual meeting in Vienna in April 2018 and was published on the IAU web pages in May 2018.

resolves

10. That, as stated in the Resolution B1 (XXIXth General Assembly, August 2015), the pursuit of the goals of the Strategic Plan: Astronomy for the Developing World should continue until the XXXI General Assembly to be held in August 2021 in Busan, Republic of Korea.

11. That, starting in 2019, the Executive Committee should prepare the implementation of the Strategic Plan 2020–2030, with particular attention to the partial overlap of the two Strategic Plans in the years 2020–2021.

12. That the goals of the Strategic Plan 2020–2030 should be pursued to their full extent starting in 2021.

13. That the Executive Committee shall include mid-term reviews about the implementation of the Strategic Plan 2020–2030 in the EC Triennial Reports to be presented to the XXXIInd and XXXIIIrd General Assemblies in 2024 and 2027.

14. That a new Strategic Plan 2030–2040 should be prepared by the Executive Committee to be presented for approval to the XXXIVth General Assembly in 2030.

RESOLUTION B1

on Geocentric and International Terrestrial Reference Systems and Frames

Proposed by the IAU Commission A2 (Rotation of the Earth)

The XXX General Assembly of International Astronomical Union,

noting

1. The essential role of the Celestial and Terrestrial Reference Systems and Frames to monitor the Earth's rotation and orientation in space with the accuracy currently required and foreseen in the near future;

2. The increasing importance of Reference Systems and Frames to science, technology, and society, and in particular to numerous astronomical and other scientific and technical activities involving precise positioning and Earth and space navigation;

3. The adoption of Resolutions B1.1 through B1.9 on reference systems by the XXIV General Assembly of the IAU in Manchester 2000, and especially of Resolution B1.3 on the definition of the Barycentric Celestial Reference System (BCRS) and Geocentric Celestial Reference System (GCRS);

4. The endorsement of IAU 2000 Resolutions B1.1 through B1.9 by Resolution 4 of the XXIII General Assembly of the International Union of Geodesy and Geophysics (IUGG) in Sapporo 2003;

5. That the GCRS is defined as a system of geocentric space-time coordinates within the framework of General Relativity with metric tensor specified by Resolution B1.3 of IAU 2000;

6. The need for a spatial reference system co-rotating with the Earth in its diurnal rotation in space for representing the Earth's orientation with respect to the GCRS;

recognizing

7. That in agreement with Resolution B1.3 of the XXIV General Assembly of the IAU in Manchester 2000, the XXIV General Assembly of the IUGG in Perugia 2007 adopted Resolution 2 that endorsed the definition of a Geocentric Terrestrial Reference System (GTRS) as a system of geocentric space-time coordinates within the framework of Gen- eral Relativity, co-rotating with the Earth, and related to the GCRS by a spatial rotation which takes into account the Earth orientation parameters;

8. That the XXIV General Assembly of the IUGG in Perugia 2007 adopted Resolution 2 that endorsed the definition of an International Terrestrial Reference System (ITRS) as the specific GTRS for which the orientation is operationally maintained in continuity with past international agreements (BIH orientation);

9. That the General Assembly of the United Nations in New York 2015 adopted Resolution 69/266 entitled "A Global Geodetic Reference Frame for Sustainable Development;"

recommends

10. That the ITRS be adopted as the preferred GTRS for scientific and technical applications; and

11. That the IAU engage, together with other concerned organizations such as the IUGG and the International Association of Geodesy, with the United Nations (UN) Global Geospatial Information Management (GGIM) Subcommittee on Geodesy in order to promote the implementation of the UN-GGIM Road Map for the Global Geodetic Reference Frame.

RESOLUTION B2

on The Third Realization of the International Celestial Reference Frame

Proposed by the IAU Working Group on the Third Realization of the International Celestial Reference Frame

The XXX General Assembly of the International Astronomical Union,

noting

1. that Resolution B2 of the XXIIIrd General Assembly (1997) resolved "that, as from 1 January 1998, the IAU celestial reference system shall be the International Celestial Reference System (ICRS)";

2. that Resolution B3 of the XXVIIth General Assembly (2009) resolved "that, as from 1 January 2010, the fundamental astrometric realization of the International Celestial Reference System (ICRS) shall be the Second Realization of the International Celestial Reference Frame (ICRF2)";

3. that Resolution B3 of the XXVIIth General Assembly (2009) resolved "that the organizations responsible for astrometric and geodetic VLBI observing programs (e.g. IERS, IVS) take appropriate measures to continue existing and develop improved VLBI observing and analysis programs to both maintain and improve ICRF2";

recognizing

4. that since the establishment of ICRF2, continued and new VLBI observing programs conducted by relevant organizations (e.g. IVS, the International VLBI Service for geodesy and astrometry) and individuals on various VLBI arrays and at multiple radio frequen- cies, have almost doubled the volume of astrometric and geodetic VLBI data collected on ICRF2 and add-on radio sources;

5. that since the establishment of ICRF2, improved instrumentation, network coverage, observation strategies, and astronomical and geophysical modeling have significantly im- proved the VLBI data quality and subsequent astrometric analysis of those data;

6. that an IAU Working Group was formed in 2012 to generate the Third Realization of the International Celestial Reference Frame using the entire astrometric and geodetic VLBI data set and state-of-the-art astronomical and geophysical modeling, with the mandate to complete that realization for presentation at the XXXth General Assembly (2018);

7. that the aforementioned Working Group has generated a prospective Third Realization of the International Celestial Reference Frame, in a coordinate frame aligned onto ICRF2, which represents a significant improvement in terms of source characterization, position accuracy and total number of sources, and thus represents a significant improve- ment in the fundamental realization of the ICRS, compared to ICRF2 adopted at the XXVIIth General Assembly (2009);

resolves

8. that, as from 1 January 2019, the fundamental realization of the International Celestial Reference System (ICRS) shall be the Third Realization of the International Celestial Reference Frame (ICRF3), as constructed by the IAU Working Group on the Third Realization of the International Celestial Reference Frame;

9. that the organizations responsible for astrometric and geodetic VLBI observing pro- grams (e.g. IVS) take appropriate measures to continue and develop such programs, at multiple radio frequencies and with a specific effort on the southern hemisphere, to both maintain and improve ICRF3;

10. that the organizations responsible for defining high-accuracy reference frames at other wavelengths take appropriate measures, together with the International Earth Rotation and Reference Systems Service (IERS), to align those reference frames onto ICRF3 with the highest possible accuracy.

RESOLUTION B3

on preservation, digitization and scientific exploration of historical astronomical data.

Proposed by IAU Inter-Division B-E Working Group on Coordination of Synoptic Observations of the Sun

The XXX General Assembly of the International Astronomical Union,

noting

1. that historical observations provide irreplaceable information regarding changes in the Sun, stars, and other objects of astronomical interest and thereby allow researchers to investigate the time domain and the nature of those transient, evolutionary or recurring changes across a far greater interval than is possible from the relatively short time-span of modern-age (born digital) observations alone;

2. that despite IAU Resolution B3 (2000), which recommended "the transfer of the historic observations onto modern media by digital techniques," the great majority of archives remain inaccessible digitally;

fearing

3. that appreciation of the unique potential which astronomy's data from the past offer, regardless of the prevailing technology, is seriously lacking;

4. that although archives and records of astronomy's analogue observations (photographic, paper, primitive magnetic tapes, etc.) are still being maintained worldwide, many are in state of increasing decay and all are at risk of loss through natural disasters and through human ignorance or error;

5. that many important datasets were acquired and curated by individual projects, which may not have resources or even plans for preserving the data much beyond the present, and

recognizing

6. that the data accumulated over the past decades and even centuries will be lost unless a concentrated action is taken to identify and preserve all significant records;

recommends

7. that a concerted effort be made to ensure the preservation, digitization, and scientific exploration of all of astronomy's historical data, both analogue and primitive digital, and associated records.

RESOLUTION B4

on a suggested renaming of the Hubble Law

Proposed by the IAU Executive Committee

The XXX General Assembly of the International Astronomical Union,

considering

1. that the discovery of the apparent recession of the galaxies, which is usually referred to as the "Hubble law," is one of the major milestones in the development of the science of Astronomy during the last 100 years and can be considered one of the founding pillars of modern Cosmology;

2. that the Belgian astronomer Georges Lemaître, in 1927 published (in French) the paper entitled *"Un Univers homogène de masse constante et de rayon croissant*

rendant compte de la vitesse radiale des nébuleuses extra-galactiques" [1]. In this he first rediscovers Friedman's dynamic solution to Einstein's general relativity equations that describes an expanding universe. He also derives that the expansion of the universe implies the spectra of distant galaxies are redshifted by an amount proportional to their distance. Finally he uses published data on the velocities and photometric distances of galaxies to derive the rate of expansion of the universe (assuming the linear relation he had found on theoretical grounds);

3. that, at the time of publication, the limited popularity of the Journal in which Lemaître's paper appeared and the language used made his remarkable discovery largely unperceived by the astronomical community;

4. that both Georges Lemaître (an IAU member since 1925 [2]) and the American astronomer Edwin Hubble (an IAU member since 1922 [3]) attended the 3rd IAU General Assembly in Leiden in July 1928 and exchanged views [4] about the relevance of the redshift vs distance observational data of the extragalactic nebulae to the emerging evolutionary model of the universe;

5. that Edwin Hubble, in 1929 published the paper entitled *"A Relation between Distance and Radial Velocity among Extra-Galactic Nebulae"* [5] in which he proposed and derived the linear distance-velocity relation for galaxies, ultimately including new velocity data in his 1931 paper with Humason [6]. Soon after the publication of his papers, the cosmic expansion became universally known as the "Hubble law";

6. that, in 1931, on invitation by the Journal Monthly Notices of the Royal Astronomical Society, G. Lemaître translated in English his original 1927 paper [7], deliberately omitting the section in which he derived the rate of expansion because he "did not find advisable to reprint the [his] provisional discussion of radial velocities which is clearly of no actual interest, and also the geometrical note, which could be replaced by a small bibliography of ancient and new papers on the subject" [8];

desiring

7. to pay tribute to both Georges Lemaître and Edwin Hubble for their fundamental contributions to the development of modern cosmology;

8. to honour the intellectual integrity of Georges Lemaître that made him value more the progress of science rather than his own visibility;

9. to highlight the role of the IAU General Assemblies in fostering exchanges of views and international discussions;

10. to inform the future scientific discourses with historical facts;

resolves

11. to recommend that from now on the expansion of the universe be referred to as the "Hubble-Lemaître law".

[1] Annales de la Société Scientifique de Bruxelles, A47, p. 49–59 (1927)

[2] Lemaître, G. 1950, Ann d' Ap., 13, 344, as translated by David L Block, 2012, in Georges Lemaître: Life, Science and Legacy, eds. R.D. Holder and S. Mitton, Astrophysics and Space Science Library, Springer-Verlag: Berlin, Vol. 395, p. 89

[3] IAU Transactions IB, 1922

[4] Humason (https://www.aip.org/history-programs/niels-bohr-library/oral-histories/4686), as reported by Sidney van den Bergh, 2011, JRASC, Vol. 105, p. 197

[5] Proceedings of the National Academy of Science, USA, 15, 168 (1929)

[6] "The velocity-distance relation among extra-galactic nebulae", Astrophysical Journal, Vol 74, p. 43–80 (1931)

[7] Monthly Notices of the Royal Astronomical Society, Vol. 91, p.483–490 (1931)

[8] Georges Lemaître, quoted by Mario Livio in Nature, Volume 479, Issue 7372, pp. 171–173

3. Report by the Resolutions Committee Chair on Resolution B4

Bruce Elmegreen, Chair of the Resolutions Committee

Electronic vote on the Resolution B4 "on a suggested renaming of the Hubble Law"

Background

Five Resolutions were proposed for approval at the XXXth IAU General Assembly (Vienna, August 20th–31st, 2018). They were announced and posted on the IAU web site on June 20th (see https://www.iau.org/news/announcements/detail/ann18029/) and initially they did not generate any comments by the members.

However, after the Resolutions were highlighted in the e-Newsletter #7 in July (see https://www.iau.org/publications/e-newsletters/html/72/), a lively discussion started on the Resolution B4 "on a suggested renaming of the Hubble Law". The comments and suggestions were well received by the Resolutions Committee, chaired by Bruce Elmegreen, and were used to improve the text of the Resolution as well as to increase the supporting bibliography.

From the comments received before the General Assembly, it was becoming clear that the opinion of the community was divided and the result of the vote would be uncertain. In such a situation, which was not necessarily foreseeable in advance, the IAU Executive Committee, in its Meeting #101 (Aug. 19th, 2018), agreed to proceed with the regular presentation of the Resolutions during the I Business Session (Aug. 21st, 2018) followed by the vote during the II Business Session (Aug. 30th, 2018), however decided to consider the vote about the Resolution B4 as indicative of the opinion of the members physically present at the II Business Session and to propose to the entire community of IAU members to express their vote electronically shortly after the GA. The result of the electronic vote would be the definitive one.

The Executive Committee decided also that the presentation of the e-vote should include the result of the straw vote at the GA and a summary of the discussion that took place both via mail and live during the II Business Session.

The final text of the Resolution B4

The text of the Resolution B4 has been modified several times from its first edition, taking into account the comments and suggestions received by the Resolution Committee up to the II Business Session of the GA. The final text which is proposed for electronic voting is reported in this Chapter.

The main motivations of the Resolution B4 are the following:

— To pay tribute to both George Lemaîre and Edwin Hubble for their fundamental contribution to the development of modern cosmology, informing future discourses with historical facts.

— To highlight the role of the IAU in fostering exchanges of views and international discussions

It should be stressed that the Resolution does not formally establish a new name of the "Hubble law," but simply suggests that in future discourses the formulation "Hubble-Lemaîre law" is preferred.

The supporting Bibliography

For convenience of the voters, the main supporting Bibliography, in particular that quoted in the Resolution, is included in this Chapter for Resolution B4.

We wish to highlight an excerpt form the paper by David L. Block, "Georges Lemaître and Stigler's Law of Eponymy," which reports an interesting comment on the matter by Lemaître himself:

"In a Comment published in Nature Mario Livio (Nature, 479, 171, 2011) has unearthed a letter from Lemaître to W. M. Smart (dated 9 March 1931). From that document, it is clear that Lemaître himself translated his 1927 paper into English and who also omitted his determination of the coefficient of expansion of the Universe (H0) from values of radial velocities available as of 1927. However, in his Comment Livio omits a vital reference, namely thoughts penned by Lemaître himself in 1950 (L'expansion de l'Univers, Bibliographie: Annales d'Astrophysique, 13, 344):

About my contribution of 1927, I do not want to discuss if I was a professional astronomer. I was, in any event, an IAU member (Cambridge, 1925), and I had studied astronomy for two years, a year with Eddington and another year in the U.S. observatories. I visited Slipher and Hubble and heard him in Washington, in 1925, making his memorable communication about the distance [to] the Andromeda nebula. While my Mathematics bibliography was seriously in default since I did not know the work of Friedmann, it is perfectly up to date from the astronomical point of view; I calculate [in my contribution] the coefficient of expansion (575 km per sec per megaparsecs, 625 with a questionable statistical correction). Of course, before the discovery and study of clusters of nebulae, there was no point to establish the Hubble law, but only to calculate its coefficient. The title of my note leaves no doubt on my intentions: A Universe with a constant mass and increasing radius as an explanation of the radial velocity of extra-galactic nebulae. I apologize that all of this is too personal. But, as noted by the author (p. 161) "the history of this science competition is not irrelevant" and it is useful to highlight the details to enable an exact understanding of the scope of the argument that can be drawn from this. (italics added for emphasis)

In 1950, Lemaître clearly did not want the rich fusion of theory and observations contained in his 1927 paper to be buried in the sands of time."

Concerning point 6 in the Resolution, we wish to highlight an interesting comment by Virginia Trimble which suggests that the expression "actual interest" used by Lemaître

8. qu'en février 2018, le projet révisé a été diffusé pour commentaires à l'ensemble des représentants des Membres nationaux de l'UAI ;

9. que la version définitive du Plan stratégique 2020-2030 a été approuvée par le Comité exécutif lors de sa 100e réunion annuelle tenue à Vienne en avril 2018 et a été publiée sur le site web de l'UAI en mai 2018 ;

recommande

10. que, comme indiqué dans la Résolution B1 (XXIXe Assemblée générale, août 2015), les objectifs du « Plan stratégique : Astronomie pour le développement » soient poursuivis jusqu'à la XXXIe Assemblée générale qui doit avoir lieu en août 2021 à Busan (République de Corée) ;

11. qu'à compter de 2019, le Comité exécutif prépare la mise en œuvre du Plan stratégique 2020-2030, en portant une attention particulière au chevauchement partiel des deux Plans stratégiques lors des années 2020-2021 ;

12. que les objectifs du Plan stratégique 2020-2030 soient poursuivis pleinement à compter de 2021 ;

13. que le Comité exécutif inclue des examens à mi-parcours quant à la mise en œuvre du Plan stratégique 2020-2030 dans ses Rapports triennaux devant être présentés lors des XXXIIe et XXXIIIe Assemblées générales en 2024 et 2027 ;

14. qu'un nouveau Plan stratégique 2030-2040 soit préparé par le Comité exécutif pour être présenté pour approbation lors de la XXXIVe Assemblée générale en 2030.

RESOLUTION B1

relative aux repères et systèmes de référence terrestre internationaux géocentriques

Proposée par la Commission A2 (Rotation de la Terre) de l'UAI

La XXXe Assemblée générale de l'Union astronomique internationale,

notant

1. le rôle essentiel des repères et systèmes de référence terrestre et céleste pour surveiller la rotation de la Terre et son orientation dans l'espace avec la précision nécessaire actuellement et prévue dans un proche avenir ;

2. l'importance croissante des repères et systèmes de référence pour la science, la technologie et la société, et en particulier pour les nombreuses activités astronomiques, scientifiques et techniques en lien avec un positionnement précis, l'observation de la Terre et la navigation spatiale ;

3. l'adoption des Résolutions B1.1 à B1.9 relatives aux systèmes de référence par la XXIVe Assemblée générale de l'UAI tenue à Manchester en 2000, et en particulier la Résolution B1.3 relative à la définition du Système de référence céleste barycentrique (BCRS) et du Système de référence céleste géocentrique (GCRS) ;

4. l'approbation des Résolutions B1.1 à B1.9 adoptées en 2000 par l'UAI, par la Résolution 4 de la XXIIIe Assemblée générale de l'Union géodésique et géophysique internationale (UGGI) à Sapporo en 2003 ;

5. que le GCRS est défini comme un système de coordonnées d'espace-temps géocentrique dans le cadre de la relativité générale avec tenseur métrique précisé par la Résolution B1.3 adoptée en 2000 par l'UAI ;

6. le besoin d'un système de référence spatial en corotation avec la Terre dans sa rotation diurne dans l'espace pour représenter l'orientation de la Terre par rapport au GCRS ;

reconnaissant

7. qu'en accord avec la Résolution B1.3 de la XXIVe Assemblée générale de l'UAI tenue à Manchester 2000, la XXIVe Assemblée générale de l'UGGI qui a eu lieu à Pérouse en 2007 a adopté sa Résolution 2 approuvant la définition d'un Système de référence terrestre géocentrique (GTRS) en tant que système de coordonnées d'espace-temps géocentrique dans le cadre de la relativité générale, en corotation avec la Terre, et lié au GCRS par une rotation spatiale qui tient compte des paramètres d'orientation de la Terre ;

8. que la XXIVe Assemblée générale de l'UGGI tenue à Pérouse en 2007 a adopté sa Résolution 2 approuvant la définition d'un Système de référence terrestre international (ITRS) en tant que GTRS spécifique dont l'orientation est entretenue de manière opérationnelle en continuité avec les accords internationaux passés (orientation BIH) ;

9. que l'Assemblée générale des Nations Unies à New York en 2015 a adopté sa Résolution 69/266 intitulée « *Repère de référence géodésique mondial pour le développement durable* » ;

recommande

10. que l'ITRS soit adopté en tant que GTRS préféré pour les applications scientifiques et techniques ;

11. que l'UAI œuvre, avec les autres organisations concernées telles que l'UGGI et l'Association internationale de géodésie, de concert avec le Sous-comité sur la gestion de l'information géospatiale à l'échelle mondiale (GGIM) des Nations Unies (ONU) en matière de géodésie afin de promouvoir la mise en œuvre de la feuille de route UN-GGIM pour le repère de référence géodésique mondial.

RESOLUTION B2

relative à la troisième version du Repère céleste international de référence

Proposée par le Groupe de travail de l'UAI sur la troisième version du Repère céleste international de référence

La XXXe Assemblée générale de l'Union astronomique internationale,

notant

1. que la Résolution B2 de la XXIIIe Assemblée générale (1997) a décidé « qu'à compter du 1er janvier 1998, le système céleste de référence de l'UAI sera le Système céleste international de référence (ICRS) » ;

2. que la Résolution B3 de la XXVIIe Assemblée générale (2009) a décidé « qu'à compter du 1er janvier 2010, la version astrométrique fondamentale du Système céleste international de référence (ICRS) sera la deuxième version (ICRF2) du Repère céleste international de référence » ;

3. que la Résolution B3 de la XXVIIe Assemblée générale (2009) a décidé « que les organisations responsables des programmes d'observations VLBI astrométriques et géodésiques (c.-à-d. l'IERS et l'IVS) devront prendre des mesures appropriées afin de poursuivre et d'améliorer les programmes d'observations et d'analyse VLBI pour maintenir et améliorer l'ICRF2 » ;

reconnaissant

4. que depuis l'introduction de l'ICRF2, la poursuite de programmes d'observations VLBI et la mise en œuvre de nouveaux programmes menés par les organisations concernées (par ex. l'IVS, le Service international VLBI pour la géodésie et l'astrométrie) et des particuliers sur différents réseaux VLBI et à des radiofréquences multiples, ont presque doublé le volume des données VLBI astrométriques et géodésiques collectées sur l'ICRF2 et des radiosources complémentaires ;

5. que depuis l'introduction de l'ICRF2, une instrumentation améliorée, la couverture réseau, les stratégies d'observation et la modélisation astronomique et géophysique ont considérablement amélioré la qualité des données VLBI et l'analyse astrométrique de ces dernières ;

6. qu'un Groupe de travail de l'UAI a été constitué en 2012 pour élaborer la troisième version du Repère céleste international de référence en utilisant l'ensemble complet des données VLBI astrométriques et géodésiques ainsi que la modélisation astronomique et géophysique la plus récente, avec pour mandat d'achever cette version en vue de sa présentation lors de la XXXᵉ Assemblée générale (2018) ;

7. que le Groupe de travail précité a élaboré un projet de troisième version du Repère céleste international de référence, dans un repère de coordonnées aligné sur celui de l'ICRF2, qui représente un progrès important en termes de caractérisation des sources, de précision de la position et de nombre total de sources, et constitue donc une amélioration notable de la version fondamentale de l'ICRS, par rapport à l'ICRF2 adopté lors de la XXVIIᵉ Assemblée générale (2009) ;

décide

8. qu'à compter du 1er janvier 2019, la version fondamentale du Système céleste international de référence (ICRS) sera la troisième version du Repère céleste international de référence (ICRF3), telle qu'elle a été élaborée par le Groupe de travail de l'UAI sur la troisième version du Repère céleste international de référence ;

9. que les organisations responsables des programmes d'observations VLBI astrométriques et géodésiques (c.-à-d. l'IVS) devront prendre des mesures appropriées afin de poursuivre et d'améliorer ces programmes, à des radiofréquences multiples et avec un effort particulier concernant l'hémisphère Sud, pour maintenir et améliorer l'ICRF3 ;

10. que les organisations responsables de la définition des repères de référence de grande précision à d'autres longueurs d'onde devront prendre des mesures appropriées, avec le Service international pour la rotation terrestre et les systèmes de référence (IERS), pour aligner ces repères de référence sur l'ICRF3 avec la plus grande précision possible.

RESOLUTION B4

visant à renommer la loi de Hubble

Proposée par le Comité exécutif de l'UAI

La XXXe Assemblée générale de l'Union astronomique internationale,

considérant

1. que la découverte de la récession apparente des galaxies, généralement désignée comme la « Loi de Hubble », est l'une des étapes clés du développement de la science de l'astronomie au cours des 100 dernières années et peut être considérée comme l'un des piliers fondateurs de la cosmologie moderne ;

2. que l'astronome belge Georges Lemaître a publié en 1927 en français l'article intitulé « *Un Univers homogène de masse constante et de rayon croissant rendant compte de la vitesse radiale des nébuleuses extragalactiques* » [1], dans lequel il a redécouvert la solution dynamique de Friedman aux équations de relativité générale d'Einstein qui décrit un Univers en expansion. Il a également établi que l'expansion de l'Univers implique que le décalage vers le rouge des spectres des galaxies lointaines est proportionnel à leur distance. Enfin, il a utilisé des données publiées sur les vitesses et les distances photométriques des galaxies pour déduire la vitesse d'expansion de l'Univers (en prenant pour hypothèse la relation linéaire qu'il avait trouvée de manière théorique) ;

3. qu'au moment de la publication, la popularité limitée du Journal dans lequel l'article de Georges Lemaître a été publié et la langue utilisée ont fait que sa découverte exceptionnelle est passée largement inaperçue de la communauté astronomique ;

4. que aussi bien Georges Lemaître (membre de l'UAI depuis 1925 [2]) que l'astronome américain Edwin Hubble (membre de l'UAI depuis 1922 [3]) ont assisté à la 3e Assemblée générale de l'UAI tenue à Leyde en juillet 1928 et ont échangé leurs points de vue [4] concernant l'importance des données d'observations quant au lien entre le décalage vers le rouge et la distance des nébuleuses extragalactiques pour le nouveau modèle de l'évolution de l'Univers ;

5. qu'Edwin Hubble, a publié en 1929 un article intitulé « *A Relation between Distance and Radial Velocity among Extra-Galactic Nebulae* » [5] dans lequel il a proposé et établi la relation linéaire distance-vitesse pour les galaxies, puis il a ajouté de nouvelles données sur la vitesse dans l'article publié en 1931 avec Humason [6]. Peu après la publication de cet article, l'expansion cosmique a été universellement connue sous le nom de « loi de Hubble » ;

6. qu'en 1931, à l'invitation du journal *Monthly Notices of the Royal Astronomical Society*, Georges Lemaître a traduit en anglais son article de 1927 [7], en omettant délibérément la partie dans laquelle il déduisait la vitesse d'expansion parce qu'il « n'a pas jugé souhaitable de renouveler la [sa] discussion provisoire sur les vitesses radiales qui ne présentait plus d'intérêt, ni la note géométrique, qui pouvait être remplacée par une brève bibliographie des documents anciens et nouveaux sur le sujet » [8] ;

souhaitant

7. rendre hommage à la fois à Georges Lemaître et à Edwin Hubble pour leurs contributions fondamentales au développement de la cosmologie moderne ;

8. honorer l'intégrité intellectuelle de Georges Lemaître qui l'a amené à accorder davantage d'importance au progrès de la science qu'à sa propre visibilité ;

9. mettre en lumière le rôle que jouent les Assemblées générales de l'UAI pour stimuler les échanges de vue et les discussions internationales ;

10. informer les discours scientifiques futurs avec des faits historiques ;

décide

11. de recommander que désormais l'expansion de l'Univers soit désignée comme la « loi de Hubble-Lemaître ».

---

[1] Annales de la Société Scientifique de Bruxelles, A47, p. 49-59 (1927)
[2] Lemaître, G. 1950, Ann d'Ap., 13, 344, tel que traduit par David L Block, 2012, dans *Georges Lemaître : Life, Science and Legacy*, eds. R.D. Holder and S. Mitton, Astrophysics and Space Science Library, Springer-Verlag: Berlin, Vol. 395, p. 89
[3] IAU Transactions Vol. 1, 1922
[4] Humason (https://www.aip.org/history-programs/niels-bohr-library/oral-histories/4686), selon indications de Sidney van den Bergh, 2011, JRASC, Vol. 105, p. 197
[5] *Proceedings of the National Academy of Science*, USA, 15, 168 (1929)
[6] « *The velocity-distance relation among extra-galactic nebulae* », Astrophysical Journal, Vol 74, p. 43-80 (1931)
[7] *Monthly Notices of the Royal Astronomical Society*, Vol. 91, p.483-490 (1931)
[8] Georges Lemaître, cité par Mario Livio dans *Nature*, Volume 479, Issue 7372, pp. 171-173 (2011)

Transactions IAU, Volume XXX
Proc. XXX IAU General Assembly, August 2018
Debra Elmegreen, ed.
doi:10.1017/S1743921323005641

CHAPTER VI

THIRTIETH GENERAL ASSEMBLY

EXECUTIVE COMMITTEE REPORT: 2015–2018

Piero Benvenuti, General Secretary

1. Introduction

The main actions of the Executive Committee during the 2015–2018 period can be summarized under the following categories:

- Improve the efficiency and transparency of the IAU Administration
- Promote the junior astronomers and improve the gender balance of the individual members
- Maintain a strong link between the Executive Committee and the IAU Divisions
- Continue and improve the support to Symposia and Regional Meetings
- Prepare the Strategic Plan for the decade 2020–2030
- Offer a "neutral" forum for discussing future large facilities (with the support of the Kavli Foundation)
- Sustain and expand the activities of the IAU Offices: OAD, OAO, OYA.
- Prepare and start the celebrations for the 100 years anniversary of the IAU

The specific activities in the different categories are described in the following sections.

2. Specific activities

2.1. *Improvements in the efficiency and transparency of the IAU Administration*

Several actions were undertaken within the IAU Office, with the support of the IAU Officers, aimed at improving the efficiency of the routine tasks of the Administration. A number of tasks were streamlined and performed using modern software tools, specifically adapted to the needs of the Office. As a natural outcome of these efforts, the IAU accounting became more transparent and easily understandable.

2.1.1. National Members Database

One of the most important task of the IAU Administration is the monitoring of the payment of the annual due by the National Members (NM). In order to properly handle the matter, the Head of Administration has to take into account the NM category, the amount of the unit due for the year and any pending arrears. Individual letters inviting the payment of the annual due (including arrears) has to be mailed to all the NM at the beginning of the year and later as a reminder if necessary. Until 2015 this chore was done by hand, taking a considerable amount of time and with the danger of miscalculation of the dues. Moreover, keeping track of the historical data was cumbersome and impractical.

account the foreseen activities for the celebrations of the IAU 100 yrs anniversary in 2019.

3.1. *Triennial Report 2015–18 and the rationale for the projection 2019–21*

The full details of the budget evolution (income and expenditures) during the triennium 2015–18 can be found in the three Annual Financial Reports that have been produced for the years 2015, 2016 and 2017 that were distributed to all National Members. In this document we only comment on the global trend of the IAU assets and a few notable features.

The trend of the last two non-GA years (2016 and 2017) has been particularly positive, with a global increment of the assets of more than 450 K€ from December 2014. The positive result is due to a number of factors: a prompter payment of the dues and arrears by the NMs, a reduction of the operational expenditures (cancellation of unused insurance and maintenance contracts, new telephone contracts, rationalization of travel expenses, etc.), donations.

The current situation will allow the IAU to provide an excellent support to the participation to the XXX GA (Vienna, August 2018), exceeding the foreseen budget for travel grants, without affecting the support to the education and outreach activities (OAD and OAO). It will also allow the support to the preparation for the 100 yrs anniversary celebration in 2019.

The increase of the travel grants for the XXX and future GAs is justified by the following reasons:

1. The special Symposium IAU S349 "Under one sky: the IAU centenary symposium" has several invited speakers that are by now retired and have little or no access to institutional funds to participate in the Symposium. Their presence is however essential for its success, therefore they should have access at least to partial travel support.

2. It was noted that, while the SOC of the non-GA Symposia may use the registration fee as well as local sponsors' funds to support invited speakers, this is not true for Symposia and Focus Meetings at the GA. At the last Officers Meeting it was therefore decided to waive the registration fee of up to 4 Symposia invited speakers to and up to 2 Focus Meetings ones.

3. In 2017 the EC decided to institute the IAU PhD Prize that rewards yearly the best PhD Thesis in each of the fields covered by the 9 IAU Divisions. The prize consists in the invitation to participate in the next GA, covering the registration fee, travel and 1 week subsistence.

The other notable items that have been considered in the preparation of the 2019–21 projection are the activities for the Celebration of the IAU 100 yrs and the introduction of the Office of Astronomy Education (OAE) which is being proposed as part of the new IAU 2020–2030 Strategic Plan.

An additional item, proposed during the Officers Meeting 2018 and now included under Scientific Activity, is the support of an annual meeting of the Junior Members.

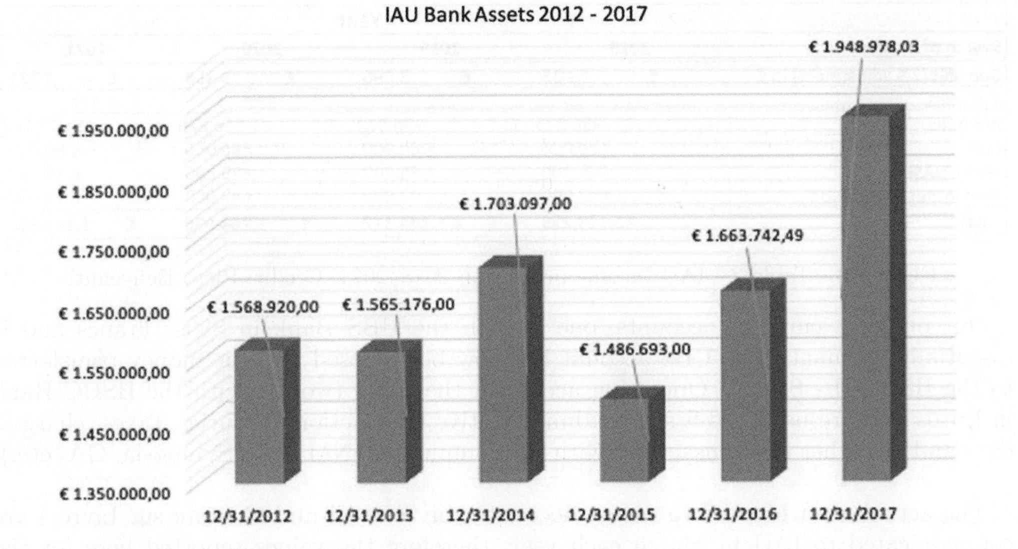

Figure 1. IAU Bank assets evolution 2012–2017. Credit: Piero Benvenuti

Overview of IAU Bank Assets 2012-2017						
CURRENT ACCOUNTS (EUR)	**12/31/2012**	**12/31/2013**	**12/31/2014**	**12/31/2015**	**12/31/2016**	**12/31/2017**
HSBC I - Administration	€ 22.190,00	€ 38.758,00	€ 52.547,00	€ 230.399,00	€ 8.588,32	€ 36.651,94
HSBC II - Scientific operations	€ 190.237,00	€ 165.821,00	€ 275.949,00	€ 228.187,00	€ 623.854,67	€ 877.987,06
USB	€ 67.504,00	€ 70.014,00	€ 69.965,00	€ 69.909,00	€ 69.851,59	€ 69.799,57
Balance Current Accounts	**€ 279.931,00**	**€ 274.593,00**	**€ 398.461,00**	**€ 528.495,00**	**€ 702.294,58**	**€ 984.438,57**
SAVING ACCOUNTS (EUR)						
HSBC Livret "A"	€ 80.475,00	€ 81.749,00	€ 82.686,00	€ 83.426,00	€ 84.052,01	€ 84.682,40
HSBC Compte sur livret	€ 508.070,00	€ 508.070,00	€ 520.801,00	€ 524.772,00	€ 527.395,90	€ 529.857,06
D.A.T. Renouvelable	€ 350.000,00	€ 350.000,00	€ 350.000,00	€ 350.000,00	€ 350.000,00	€ 350.000,00
HSBC Trésorerie Plus	€ 350.444,00	€ 350.764,00	€ 351.149,00	€ -		
Balance Saving Accounts	**€ 1.288.989,00**	**€ 1.290.583,00**	**€ 1.304.636,00**	**€ 958.198,00**	**€ 961.447,91**	**€ 964.539,46**
TOTAL ASSESTS	**€ 1.568.920,00**	**€ 1.565.176,00**	**€ 1.703.097,00**	**€ 1.486.693,00**	**€ 1.663.742,49**	**€ 1.948.978,03**

Figure 2. Table 1 - IAU Bank assets evolution 2012–2017. Credit: Piero Benvenuti

The general rationale that has been followed in preparing the projection was to contain the operational expenditures (EC and Administration) to the current level and to invest the accumulated profit in Scientific, Educational and Outreach activities with the objective to reach, at the end of the triennium in 2021, an asset equal to about 1 year income.

As far as the income is concerned, it is proposed to maintain the 2% annual adjustment of the unit contribution to take into account the average index of inflation. The same adjustment has been recently proposed by the ISC (ex-ICSU) and approved at large majority by the ICSU affiliated Unions.

3.2. *Detailed Budget Projection*

3.2.1. Bank assets

The IAU holds and operates 3 current bank accounts and 3 saving accounts.

	Year			
Source of income	2018	2019	2020	2021
Due unit (2% adjustment) in €	€ 3.035	€ 3.096	€ 3.158	€ 3.221
# of units	315	322	322	322
NM dues	€ 956.025	€ 996.912	€ 1.016.876	€ 1.037.162
CUP	€80.000	€ 80.000	€80.000	€ 80.000
KAVLI/NASL	€25.200	€ 25.200	€25.200	€ 25.200
The Gruber Foundation	€42.000	€ 42.000	€42.000	€ 42.000
Total	**€1.103.225**	**€ 1.144.112**	**€ 1.164.076**	**€ 1.184.362**

Figure 3. Table 2 - IAU Income projection 2018–2021. Credit: Piero Benvenuti

Out of the 3 currents accounts, one is with the UBS Bank in Swiss Francs and is essentially dormant (The UBS account has now been closed and the money transferred to the HSBC-II - Science Operation account), the other two are with the HSBC Bank in Euros and are used the first for administrative transactions (salaries, taxes, charges, etc.) and the other for transactions with the community (NM dues, Symposia, GA, etc.).

The actual asset of the two saving accounts (Livret "A" and Compte sur Livret) are communicated to IAU in March each year, therefore the values reported here for the year 2017 are the same as those at the end of 2016. The total asset for 2017 will have to be adjusted when the accrued interests will be known. The interests of the D.A.T Renouvelable (350.000 €) are paid every 6 months and are automatically registered in the Administration account.

The Assets data from 2015 to 2016 are reported in here in table and graphical form.

3.2.2. Sources of income

The income forecast for the years 2018–2021 is presented in Table 2. It takes into account the proposed 2% adjustment of the dues unit and the increase of the number of units associated to the admission of 4 new National Members in Category I. The CUP royalties is a prudent estimate and the KAVLI/NASL for the ISYAs (International Schools for Young Astronomers) and Gruber for the TGF (The Gruber Fellowship) are funds in transit.

3.2.3. Science expenditure projection

The expenditure forecast in the area of Science support for the years 2018–2021 is reported in Table 3. The main variation from the previous triennial budgets is the increase of the grants allocation for the XXX and XXXI GAs.

The rationale for the increase has been described in Section 3.1.

3.2.4. Education and Outreach expenditure projection

The expenditure forecast in the area of Education and Outreach support for the years 2018–2021 is reported in Table 4. The main variations from the previous triennial budgets are:

An increase of the allocation for the ISYA Schools to 50 K€ per year. This will allow the OYA to accept two ISYA per year instead of the current number of 3 schools every two years. Assuming that the KAVLI/NASL annual contribution remains at the level of $30 K, this implies an increase of the IAU annual contribution of 15 K€. Since the cost

SCIENTIFIC EXPENDITURE - PROJECTION 2018-2021								
			2017	*2018*	2019	2020	2021	Total (triennium)
M	General Assembly							
	M1	Preparation and organization cost		€ 30.000	€ 5.000	€ 10.000	€ 30.000	
	M2	GA Grants		€ 550.000	€ -	€ -	€ 450.000	
	M3	Other		€ 5.000	€ -	€ -	€ 5.000	
M	Total		€ -	€ *585.000*	€ 5.000	€ 10.000	€ 485.000	€ 500.000
N	Scientific Activities outside the GA							
N1	Sponsored Meetings							
	N1.1	IAU Symposia outside GA		€ 60.000	€ 180.000	€ 180.000	€ 60.000	
	N1.2	Regional IAU Meetings (RIM)		€ -	€ 40.000	€ 20.000	€ -	
	N1.3	Junior Members Meetings			€ 30.000	€ 30.000	€ 30.000	
	N1.4	Cosponsored Meetings		€ 15.000	€ 15.000	€ 15.000	€ 15.000	
	Sub-total Sponsored Meetings		€ *238.479*	€ *75.000*	€ 265.000	€ 245.000	€ 105.000	€ 615.000
N2	Working Groups							
	N2.1	EC Working Groups		€ 5.000	€ 10.000	€ 10.000	€ 5.000	
		Junior Members 'meeting						
	N2.1.2	Others		€ -	€ 5.000	€ 5.000	€ -	
	N2.2	Commissions Working Groups		€ 5.000	€ 5.000	€ 5.000	€ 5.000	
	N2.2.1	Astronomy & UNESCO World Heritage		€ 5.000	€ 5.000	€ 5.000	€ 5.000	
	Sub-total Working Groups		€ *15.855*	€ *15.000*	€ 25.000	€ 25.000	€ 15.000	€ 65.000
N	Total		€ *254.334*	€ 90.000	€ 290.000	€ 270.000	€ 120.000	€ 680.000
TOTAL SCIENTIFIC ACTIVITIES			€ *254.334*	€ 675.000	€ 295.000	€ 280.000	€ 605.000	€ 1.180.000

Figure 4. Table 3 - Scientific expenditure projection 2018–2021. Credit: Piero Benvenuti

of individual ISYA may vary, depending on the number of factors, the OYA triennial budget is reserved exclusively for the ISYA, therefore the annual positive or negative balances are carried over into the next annual budget.

The allocation of 110 K€ for the annual call of OAD Projects is raised to 120 €(the call is largely oversubscribed).

The cost of the CAP (Communicating Astronomy to the Public) Journal takes into account the transfer of responsibility from ESO to OAO.

The establishment of the Office of Astronomy Education (OAE) is proposed in the new IAU Strategic Plan 2020–2030. In the current forecast, the expenditure for the OAE is just a ramping up estimate which will have to be adjusted once the International Call for the establishment of the Office will be known.

The cost of the initiatives for the celebration of the IAU 100 years anniversary in 2019 has been spread among the 3 years 2017–2019.

The funds in transit in support of education by The Gruber Foundations remain unchanged.

The cooperation with other Unions shows a reduction of the annual contribution to ICSU (from 2020 to the new International Council for Science, ISC, which results from the merging of ICSU and ISSC that was approved in October 2017). As a matter of fact, the annual contribution to the new ISC has not yet been determined, however the EC concurred that the current policy adopted by ICSU to determine the annual due of the affiliated Union, exclusively based on the gross income of the Union, is grossly unfair, because it considers the Unions as a sort of profit enterprises and does not take into account the use of the income which, in the IAU case, is mainly devoted to supporting

			2017	2018	2019	2020	2021	Total (triennium)
EDUCATION AND OUTREACH EXPENDITURE - PROJECTION 2018-2021								
O	**Educational Activities**							
O1	Schools and Training							
	O1.1	OYA - ISYA	€ 24.794	€ 50.000	€ 50.000	€ 50.000	€ 50.000	
	O1.2	Co-sponsored Schools	€ -	€ 5.000	€ 5.000	€ 5.000	€ 5.000	
	O1.3	Others	€ -	€ 1.000	€ 1.000	€ 1.000	€ 1.000	
	Sub-Total Schools and Training		€ 24.794	€ 56.000	€ 56.000	€ 56.000	€ 56.000	€ 168.000
O2	**OAD**							
	O2.1	OAD Operations	€ 70.000	€ 70.000	€ 70.000	€ 70.000	€ 70.000	
	O2.2	OAD Projects	€ 124.459	€ 150.000	€ 150.000	€ 150.000	€ 150.000	
	O2.3	Start-up Contribution to ROADS	€ -	€ -	€ 5.000	€ -	€ 5.000	
	O2.4	OAD Steering Committee	€ 1.555	€ 5.000	€ 5.000	€ 5.000	€ 5.000	
	O2.5	OAD Reviews	€ -	€ -	€ -	€ 10.000	€ -	
	Sub-Total OAD		€ 196.014	€ 225.000	€ 230.000	€ 235.000	€ 230.000	€ 695.000
O3	**OAO**							
	O3.1	IAU Contribution to OAO Operation	€ 50.000	€ 50.000	€ 50.000	€ 50.000	€ 50.000	
	O3.2	External Contributions to OAO	€ 18.069	€ -	€ -	€ -	€ -	
	O3.4	CAP Journal (editing)	€ 3.978	€ 3.280	€ 3.300	€ 3.300	€ 3.300	
	O3.5	CAP Journal (printing & distribution)	€ 2.220	€ 7.000	€ 7.200	€ 7.200	€ 7.200	
	O3.5	CAP Conference	€ -	€ 5.000	€ -	€ -	€ 5.000	
	O3.6	OAO Reviews	€ -	€ -	€ -	€ 5.000	€ -	
	Sub-Total OAO		€ 74.267	€ 65.280	€ 60.500	€ 65.500	€ 65.500	€ 191.500
O4	**OAE**							
	O4.1	IAU Contribution to OAE Operations	€ -	€ -	€ 10.000	€ 20.000	€ 20.000	
	O4.2	Travels	€ -	€ -	€ 5.000	€ 15.000	€ 15.000	
	O4.3	Projetcs	€ -	€ -	€ 5.000	€ 20.000	€ 20.000	
	Sub-Total OAE		€ -	€ -	€ 20.000	€ 55.000	€ 55.000	€ 130.000
	Total Educational Activities		€ 295.075	€ 346.280	€ 366.500	€ 411.500	€ 406.500	€ 1.184.500
O5	**IAU 100yrs**							
	O4.1	Travel	€ 13.027	€ 11.500	€ 55.000	€ -	€ -	
	O4.2	Outsourcing	€ -	€ 40.000	€ 50.000	€ -	€ -	
	O4.3	Projects	€ -	€ 23.000	€ 50.000	€ -	€ -	
	O4.4	Production	€ 10.816	€ 25.000	€ 40.000	€ -	€ -	
	Total IAU 100yrs		€ 23.842	€ 99.500	€ 195.000	€ -	€ -	€ 195.000
P	**Education and Research Support (Funds in Transit)**		2017	2018	2019	2020	2021	Total (triennium)
P1	The Gruber Foundation							
	P1.1	TGF Fellowship	€ 41.750	€ 41.750	€ 41.750	€ 41.750	€ 41.750	
	P1.2	TGF Support (GA)	€ -	€ 7.000	€ -	€ -	€ 7.000	
	Total Support - Funds in transit		€ 41.750	€ 48.750	€ 41.750	€ 41.750	€ 48.750	€ 132.250
Q	Cooperation with Other Unions							
Q1	Delegates Travels		€ 615	€ 3.000	€ 3.000	€ 3.000	€ 3.000	
Q2	Dues to Other Unions							
	Q2.1	ICSU	€ 22.489	€ 22.489	€ 22.939	€ 23.398	€ 23.866	
	Q2.1.1	ICSU Women	€ 2.590	€ 5.000	€ 7.500	€ -	€ -	
	Q2.2	IUCAF	€ 5.000	€ 5.000	€ 5.000	€ 5.000	€ 5.000	
	Q2.3	COSPAR	€ 5.000	€ 5.000	€ 5.000	€ 5.000	€ 5.000	
	Q2.4	Others (ICSTI)	€ -	€ 800	€ 800	€ 800	€ 800	
	Total Cooperation with Other Unions		€ 35.694	€ 41.289	€ 44.239	€ 37.198	€ 37.666	€ 119.102
TOTAL EDUCATION AND OUTREACH			€ 396.362	€ 535.819	€ 647.489	€ 490.448	€ 492.916	€ 1.630.852

Figure 5. Table 4 - Education and outreach expenditure projection. Credit: Piero Benvenuti

Science, Education and Outreach. On this basis, the EC authorized the GS to communicate to ICSU a unilateral reduction of the annual due of 50%, in view of a hopefully more rationale due policy of the new ISC.

3.2.5. Executive Committee and Publications expenditure projection

The expenditure projection in the area of Executive Committee and Publications for the years 2018–2021 is reported in Table 5. There are no major variations from the previous triennial budgets. It should be noted that the budget for EC expenses takes into account the two EC Meetings in the GA years and the fact that the Division Presidents are invited to attend one EC Meeting between GAs.

EXECUTIVE COMMITTEE AND PUBLICATIONS EXPENDITURE - PROJECTION 2018-2021								
			2017	2018	2019	2020	2021	Total (triennium)
R	Executive Committee & Officers							
R1	Executive Committee Meetings							
	R1.1	EC Expenses (others)	€ 26.639	€ 100.000	€ 50.000	€ 35.000	€ 100.000	€ 185.000
R2	Officers Meetings		€ 12.650	€ 15.000	€ 15.000	€ 15.000	€ 15.000	€ 45.000
R3	Officers Expenditures (Other)							
	R3.1	General Secretary						
	R3.1.1	GS Paris Duty	€ 20.569	€ 20.000	€ 20.000	€ 20.000	€ 20.000	
	R3.1.2	GS Other Esxpenses	€ 14.706	€ 8.000	€ 12.000	€ 12.000	€ 8.000	
	R3.1.3	Past GS Expenses	€ -	€ 1.500	€ 3.000	€ 3.000	€ 1.500	
	R3.2	AGS Expenses	€ 4.305	€ 2.000	€ 2.000	€ 2.000	€ 2.000	
	R3.3	President Expenses	€ 896	€ 2.000	€ 3.000	€ 3.000	€ 1.500	
	R3.3.1	President Elect Expenses	€ -	€ 2.000	€ 2.000	€ 2.000	€ 1.000	
	R3.3.2	Past President Expenses	€ 1.197	€ 1.000	€ 1.000	€ 1.000	€ 1.000	
	Total Executive Committee & Officers		€ 80.963	€ 151.500	€ 108.000	€ 93.000	€ 150.000	€ 351.000
R4	IAU Press Office		€ 646	€ 12.000	€ 500	€ 500	€ 12.000	€ 13.000
S	Publications							
	S1	Information Bulletin	€ -	€ 5.000	€ 5.000	€ 5.000	€ 5.000	€ 15.000
TOTAL EXECUTIVE COMMITTEE			€ 81.609	€ 168.500	€ 113.500	€ 98.500	€ 167.000	€ 379.000

Figure 6. Table 5 - Executive Committee and Publications expenditure projection. Credit: Piero Benvenuti

3.2.6. Administration and Operations expenditure projection

The expenditure projection in the area of Administration and Operations for the years 2018–2021 is reported in Table 7. The forecast shows some economies due to the end of contract of the part-time archivist and to the rationalization of the general office expenditures.

3.3. *Summary*

The summary of the budget projection for the years 2018–2021 is reported in Table 7. As indicated in Section 2, the budget has been designed to arrive at the end of 2021 with an asset roughly equivalent to one-year income. In order to achieve that goal, the net results of the years 2018, 2019 and 2021 are negative, while it is positive for the "intermediate" year 2020. Considering that, as explained in Section 2, this triennium is somewhat special due to the 100 years anniversary celebrations, the next triennium 2022–2024 should resume the usual trend, i.e. a positive result in the two non-GA years and negative in the GA year.

The evolution of the Bank assets is reported in Figure 8 that clearly shows the gradual reduction of the assets to arrive at the end of 2021 with an amount of $\sim$1.084 K€, equivalent to the average income in the triennium:$\sim$ 1.105 K€.

ADMINISTRATION AND OPERATIONS EXPENDITURE - PROJECTION 2018-2021								
			2017	2018	2019	2020	2021	Total (triennium)
T	Secretariat & Administration							
T1	Gross staff salaries including taxes & charges		€ 113.533	€ 104.400	€ 104.400	€ 104.400	€ 104.400	
	T1.1	Extra Charges on salaries	€ 48.658	€ 40.000	€ 40.000	€ 40.000	€ 40.000	
	T1.2	End of contract indemnities	€ 27.522	€ -	€ -	€ -	€ -	
	Sub-total Salaries & Charges		**€ 189.713**	**€ 144.400**	**€ 144.400**	**€ 144.400**	**€ 144.400**	**€ 433.200**
T2	Staff training and other expenses		€ -	€ -	€ 1.000,00	€ 1.000,00	€ 1.000,00	
	T2.1	Secretariat travel	€ 2.487	€ 1.000,00	€ 1.000,00	€ 1.000,00	€ 1.000,00	
	Sub-total Staff Training		**€ 2.487**	**€ 1.000**	**€ 2.000**	**€ 2.000**	**€ 2.000**	**€ 6.000**
T3	Outsourced tasks							
(1)	T3.1	Web support and development (ESO)	€ 20.000	€ 20.000	€ 20.000	€ 20.000	€ 20.000	
	T3.2	Data Base management (ESO)	€ 26.581	€ 27.000	€ 27.000	€ 27.000	€ 27.000	
	T3.3	IT support from IAP	€ 1.800	€ 1.800	€ 1.800	€ 1.800	€ 1.800	
	T3.4	Personnel administration (payslips, etc.)	-€ 3.336	€ 2.000	€ 2.000	€ 2.000	€ 2.000	
(2)	T3.5.1	Accounting	€ -	€ 2.000	€ 2.000	€ 2.000	€ 2.000	
	T3.5.2	SAGE/CIEL subscription	€ 771	€ 800	€ 800	€ 800	€ 800	
	T3.6	External Auditing	€ 4.896	€ 5.000	€ 5.000	€ 5.000	€ 5.000	
(4)	T3.7	Legal support	€ 2.533	€ -	€ 1.000	€ 1.000	€ 1.000	
	T3.8	Mi-Voice e-voting subscription	€ 2.160	€ 2.800	€ 2.000	€ 2.000	€ 2.800	
	Sub-total Outsourced tasks		**€ 55.405**	**€ 61.400**	**€ 61.600**	**€ 61.600**	**€ 62.400**	**€ 185.600**
T4	General Office Expenditure							
	T4.1	Post	€ 435	€ 500	€ 500	€ 500	€ 500	
	T4.2	Telephone and Internet	€ 758	€ 800	€ 800	€ 800	€ 800	
	T4.3	Rent CNRS/IAP, Paris	€ 4.403	€ 4.500	€ 4.500	€ 4.500	€ 4.500	
	T4.3.1	Office Maintenance	€ 3.239	€ 500	€ 500	€ 1.500	€ 500	
	T4.4	IT Software & Hardware	€ 3.365	€ 1.500	€ 1.500	€ 2.500	€ 1.500	
	T4.5	Printer/photocopier (rent and consumables)	€ 4.096	€ 2.400	€ 2.400	€ 2.400	€ 2.400	
	T4.6	Office consumables	€ 691	€ 500	€ 500	€ 500	€ 500	
	T4.7	Miscellaneous	€ 307	€ 300	€ 300	€ 300	€ 300	
	T4.8	Bank charges	€ 4.331	€ 3.500	€ 3.500	€ 3.500	€ 3.500	
	Sub-total General Office		**€ 21.625**	**€ 14.500**	**€ 14.500**	**€ 16.500**	**€ 14.500**	**€ 45.500**
T5	Entertainement expense, gifts and tips							
	T5.1	Restaurants and catering	€ 113	€ 150	€ 150	€ 150	€ 150	
	T5.2	Contributions (IAP fête des enfants) and tips	€ 81	€ 300	€ 300	€ 300	€ 300	
	Sub-total Gifts and tips		**€ 195**	**€ 450**	**€ 450**	**€ 450**	**€ 450**	**€ 1.350**
U	Taxes on Associations' revenues		€ 696	€ 1.000	€ 1.000	€ 1.000	€ 1.000	
TOTAL ADMINISTRATION AND OPERATIONS			**€ 270.120**	**€ 222.750**	**€ 223.950**	**€ 225.950**	**€ 224.750**	**€ 674.650**

Figure 7. Table 6 - Administration and Operations expenditure projection. Credit: Piero Benvenuti

IAU budget projection for the triennium 2019-2021	Closed	Projection	Triennium 2019-2021			Total
	2017	2018	2019	2020	2021	2019-2021
Income	€ 1.268.154	€ 1.103.225	€ 1.144.112	€ 1.164.076	€ 1.184.362	€ 3.492.550
Science	€ 254.334	€ 675.000	€ 265.000	€ 250.000	€ 575.000	€ 1.090.000
Education & Outreach	€ 396.362	€ 535.819	€ 647.489	€ 490.448	€ 492.916	€ 1.630.852
Executive Committee	€ 81.609	€ 168.500	€ 113.500	€ 98.500	€ 167.000	€ 379.000
Administration & Operations	€ 247.300	€ 222.750	€ 223.950	€ 225.950	€ 224.750	€ 674.650
Total expenditure	€ 979.604	€ 1.602.069	€ 1.249.939	€ 1.064.898	€ 1.459.666	€ 3.774.502
Net result	€ 288.549	-€ 498.844	-€ 105.827	€ 99.178	-€ 275.304	-€ 281.952

Figure 8. Table 7 - Summary of the budget projection 2018–2021. Credit: Piero Benvenuti

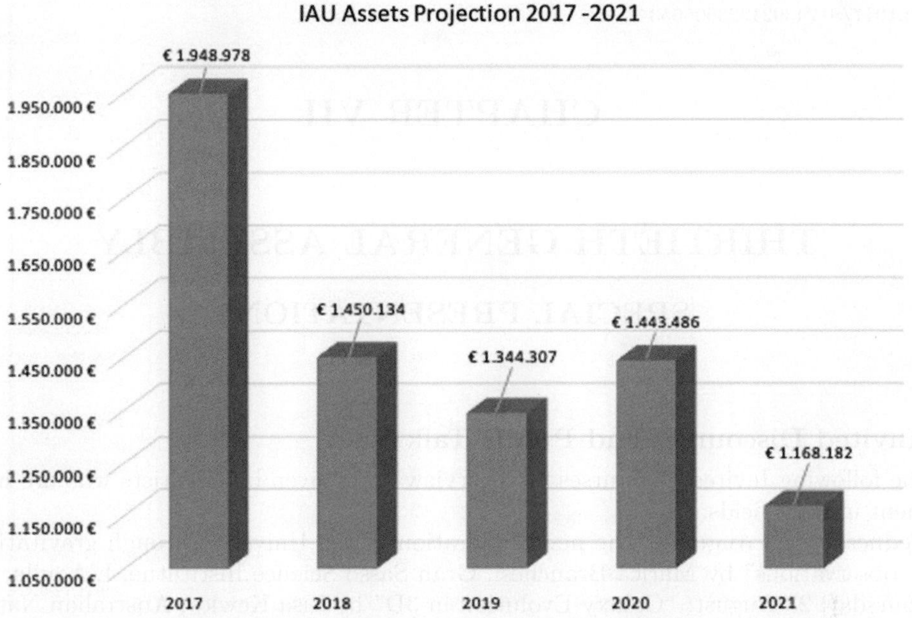

Figure 9. IAU assets projection 2017–2021. Credit: Piero Benvenuti

Transactions IAU, Volume XXX
Proc. XXX IAU General Assembly, August 2018
Debra Elmegreen, ed.
doi:10.1017/S1743921323005653

CHAPTER VII

THIRTIETH GENERAL ASSEMBLY

SPECIAL PRESENTATIONS

1. Invited Discourses and Public Talk

The following Invited Discourses are overview talks given by scientists who are highly eminent in their fields.

Wednesday, 22 August: "The new exploration of the Universe through gravitational-wave observations" by Marica Branchesi, Gran Sasso Science Institutue, L'Aquila, Italy

Thursday, 23 August: "Galaxy Evolution in 3D" by Lisa Kewley, Australian National University, Weston Creek ACT, Australia

Monday, 27 August: "Observing Planet Formation" by Sean Andrews, Harvard-Smithsonian Center for Astrophysics, Cambridge, Massachusetts, USA

On August 24, Robert Williams, Past President from the Space Telescope Science Institute and Distinguished Osterbrock Professor of the University of California Santa Cruz, USA, gave a public lecture on "Probing the Distant Universe with the Hubble Space Telescope."

2. Launch of IAU 100th Anniversary Celebration

The inaugural celebration of 100 years of the IAU took place on Tuesday August 28, with music, an IAU video, and new ideas and expectations presented by Luiz Felipe Santiago Rodriguez and Olesky Golubov. Past President Catherine Cesarsky presented her "Reflections on 100 years of IAU."

Note added in 2023: Prof. Cesarsky's full speech, with photos, was subsequently published in "Under One Sky: The IAU Centenary Symposium," Proceedings of the International Astronomical Union, Volume 349, pp. 25–47, Cambridge University Press, 2019, available online, so is not repeated here.

Dr. Ewine van Dishoeck, IAU President-Elect and Chair of IAU100 Task Force:

In 2019, the IAU will celebrate its 100 year anniversary. To commemorate this milestone, the IAU will organize a year-long celebration to highlight a century of astronomical discoveries and to further stimulate the use of astronomy as a tool for education, development and diplomacy. The IAU100 theme "Under One Sky" was chosen to emphasize the worldwide nature of our field and the role that the IAU has played in bringing people together and uniting them to explore the universe.

Planning of IAU100 started in 2017. It is appropriate that the kick-off of these celebrations takes place here at the 2018 IAU General Assembly in Vienna. Some of you already attended the opening of the IAU100 "Above and Beyond" Exhibition on Monday, featuring also a delicious huge IAU100 Sacher-Torte. Let me give you a brief overview of what we are planning to do in the coming year. In short, a lot! My thanks go to the IAU Task Force and secretariat, whose names are listed below, and in particular to our IAU100 coordinator, Jorge Rivero González, as the main driver of all the activities. See our website `www.iau-100.org` for up-to-date information on everything that will be happening.

The IAU100 activities will take place at global and regional levels, and especially at the national and local levels. We are preparing a comprehensive program of flagship initiatives to reach audiences worldwide through the IAU National Outreach Coordinator contact points in 123 countries. The target audiences range from the global astronomical community, national science organizations and societies, to policy-makers, students, amateur astronomers as well as families and the general public. Its program and organization make good use of the experience developed during the International Year of Astronomy 2009 (IYA2009), but IAU100 has a slightly different focus that includes highlighting astronomical discoveries over the past century as well as the role of the IAU in making them happen.

To address the latter point: there will be two IAU100 books that will both appear in 2019:

- The International Astronomical Union: Uniting the Community for 100 Years, by Johannes Andersen, David Baneke and Claus Madsen (Springer)
- IAU Symposium 349 "Under One Sky": the IAU Centenary Symposium proceedings, to be edited by Chris Sterken, John Hearnshaw and David Valls-Gabaud (CUP)

Please take the opportunity to attend IAU Symposium 349 here in Vienna in the coming days to learn more about the IAU history.

The overall goals of IAU100 are:

1. Increase awareness of progress and excitement in astronomy over the past century, in particular:
   - The importance of collaborative enterprise of astronomy as a whole
   - The importance of technology development for astronomical progress
   - The coordinating role of the IAU for the global astronomical community
2. Promote widespread access to astronomy knowledge and observing experiences.
3. Support and improve use of astronomy as a tool for education, development and diplomacy.
4. Support and improve an inclusive, equalitarian and diverse astronomy community.
5. Facilitate the preservation and protection of the world's cultural and natural heritage of dark and quiet skies.
6. Raise awareness and discuss prospective new exciting developments in the next 100 years of astronomy.

Note that these goals match well with the new IAU Strategic Plan 2020–2030 approved here at the GA in Vienna.

The IAU100 celebrations are structured around a number of flagship programs. These are global programs of activities centered on specific themes that will help to achieve IAU100's main goals. They include:

- IAU100 Celebrations

Through the organization of high-level events in different countries, the IAU100 celebrations will bring together high-level representatives, policy-makers and astronomers with an exciting program centered around Goal 1. The main event will be the "IAU100 Flagship event at Palais des Academies in Brussels (Belgium)" to be organized in April 2019 at the location where the IAU was founded in 1919.

• IAU100 Exhibitions

The IAU "Above and Beyond" exhibition, developed in collaboration with Science Now, showcases major achievements in astronomy in the last century, their wider technological and cultural implications, as well as selected IAU milestones. The inauguration of this travelling exhibition happened on August 20 2018 here in Vienna. By making the entire content open source, it can be easily reproduced (and translated) for venues elsewhere in the world. Also, a low-cost version is available worldwide in the form of A0 posters.

• New Worlds: Are We Alone?

The actions of this flagship program will foster a sense of global citizenship and critical thinking by encouraging the public at large to think about our place in the universe through the organization of activities related to exoplanets and astrobiology. Most notably, following previous IAU experience, the "NameExoWorld" project will offer the possibility to all nations in the world the possibility to give a name to an exoplanet system.

• 100 Years of General Relativity: Solar Eclipse

The year 2019 also marks the 100th anniversary of the solar eclipse observations that served as the first successful test of Einstein's Theory of General Relativity that fundamentally changed the way we understand gravity and the universe. This flagship program will highlight and explain this important milestone by organizing commemorative actions as well as reaching out to larger audiences, including schools, to raise awareness about the importance of gravity and Einstein's Theory of General Relativity by building on recent discoveries such as the detections of gravitational waves. A highlight will be the events, conferences and educational activities at Principe Island in late May 2019 where Eddington's expedition observed the Solar eclipse of 29 May 1919. Also, throughout the year, the "Einstein Schools" will educate children worldwide about the role of gravity in shaping our universe, with STEM professionals acting as mentors and encouraging enrollment.

• Astronomy for All

One important focus of the IAU100 celebrations will be the implementation of actions to support and encourage women and minority scientists and engineers at all career levels and promote inclusive, equitable work environments within astronomy-related careers. In addition, this flagship program will focus on facilitating access to astronomical resources and careers for people with special educational or physical needs, or those who might be excluded for their particular race or gender.

One highlight is the itinerant "Inspiring Stars" exhibition, which premiered here at the GA in Vienna a few days ago, and which focuses on visually challenged children and adults. Please go and see the exhibition if you have not yet done so. Another planned activity is the IAU "Women and Girls in Astronomy" day to commemorate the United Nations International Day of Women and Girls in Science on February 11th. Coordinated actions will be organized through the IAU's National Outreach Coordinator network to support women in astronomy and encourage girls for a career in science by providing access to excellent role models and mentors. Finally, the IAU will organize its first symposium ever on "Astronomy for Equity, Diversity and Inclusion – a roadmap to action within the framework of the IAU centennial anniversary": IAU Symposium 358 will be held in Japan in November 2019.

- Star Parties

Through the organization of side-walk astronomy events, IAU100 plans to enable as many people as possible, especially children, to look at the sky through a telescope and gain a basic understanding of the universe. Selected worldwide actions include the "100 Hours of Astronomy" in January 2019, which will invite amateur astronomers, educators, professional astronomers, planetaria, science centers and more to arrange astronomy-related events around the world during this 4-day period to reach out to the public at large.

Moreover, 2019 is a special year because of the 50th anniversary of the Moon landing. Therefore, on July 20 2019, we will plan "Moon landing 50 years and Sea of Tranquility sidewalk events" calling on those who own portable telescopes around the world, to show the Moon from public squares and identify the site of the historic moon landing.

- Dark Skies for All

This flagship focuses on the promotion of the preservation of the dark and quiet skies, a topic that is becoming more and more urgent for our field and that has been on IAU's agenda for a long time. Note that there are various sessions organized here at the GA that we urge you to attend. As part of IAU100, a high-level conference with UNESCO is being planned in 2019–2020. Also, a new IAU Dark Skies Ambassadors Network for public engagement on light pollution protection actions will be formed, including government lobbying actions. It will capitalise on existing educational programs on light pollution such as the Quality Lighting Teaching Kit.

I hope that this brief overview of IAU100 provides you with a "taste" of what is to come in 2019. If you want to participate, please contact your National Outreach Coordinator. You can implement one of the endorsed activities locally, or develop an activity under one of the flagship themes, or implement your own idea. Your involvement will be vital for the IAU100 success, we need you!

The IAU Task Force consists of E.F. van Dishoeck (chair), P. Benvenuti, L. Canas, S.-L. Cheung, L. Christensen, R. d'Antonio, D. Elmegreen, J. Rivero González, K. Govender, M.T. Lago, S. Torres-Peimbert, I. Robson, P. Russo, R. Williams. The IAU100 Secretariat is located at Leiden Observatory, the Netherlands.

Note added in 2023: this report reflects the presentation that was made at the IAU GA in Vienna in August 2018, at the very start of IAU100. The full report of IAU100, published in 2020 after IAU100 ended, can be found at the IAU and IAU100 websites, at `https://www.iau.org/static/archives/announcements/pdf/iau100-final-report-ann20019.pdf`

3. City Hall Reception by the Mayor

The Mayor and Governor of Vienna, Michael Ludwig, welcomes the IAU General Assembly participants at the City Hall on Tuesday evening, August 21.

Dr. Silvia Torres-Peimbert, IAU President:

Distinguished hosts and friends,

For the International Astronomical Union it has been a privilege to have been hosted by the City of Vienna at its very fine Convention Center to hold its XXX General

Assembly. More than 3000 participants of 89 different countries have registered to exchange ideas and reinforce international collaborations.

We all recognize that is an old city rich in history and culture. The City of Vienna has given us the opportunity that in addition to our daily scientific work, to enjoy the beauty of its buildings and parks, and to admire and enjoy its many museums and works of art.

In the personal side, Vienna has for me the extraordinary significance of having been the home of such distinguished musicians as Ludwig van Beethoven, Amadeus Mozart, and Johann Strauss, to extraordinary artists like Gustav Klimt, Oskar Kokoschka and Egon Schiele, whom I admire. and the father of psychoanalysis, Sigmund Freud, who has helped us to understand human behavior. In addition to those well-known valuable figures, all of the astronomers are aware that this city nurtured great scientists. I will refer only to those physicists whose research contributed to the understanding of my area of study in Astrophysics. I wish to refer to some of these distinguished physicists, Ernst Mach (1838 - 1916) for his study on shock waves, Ludwig Boltzmann (1844 - 1906) for the correlation of thermodynamics and mechanics, Lise Meitner (1878 - 1968) whose research in the field of radioactivity eventually led to the discovery of nuclear fission of uranium, Erwin Schrödinger (1887 - 1961), for the development of the atomic theory, Victor Hess (1883 – 1964) for the discovery of Cosmic Rays. And finally, I cannot resist commenting about another special attraction in this city: the famous Viennese coffee houses that have been included by UNESCO in the national inventory of intangible cultural heritage: "Viennese coffee house culture" with their delicious Sachertorte and Apple Strudel.

The fascination of science, and in particular of astronomy, is continuously expanding in most countries, among adults and children. I believe that this is marvelous since it shows that human curiosity and respect for nature and knowledge is an unperishable value.

4. Inspiring Stars Inaugural Event

The inaugural exhibition of "Inspiring Stars" occurred on August 22 at the Austria Center Vienna. "Inspiring Stars" is a travelling international exhibition designed to highlight, support, and promote inclusive initiatives in outreach and teaching, and at a professional level, in the field of astronomy. The IAU invited all attending members, members of Austrian organisations that cater for people with disabilities and/or impairments, the Austrian astronomical community, and members of the public to visit the exhibition and the special events taking place during the inauguration. With this exhibit, the IAU aims to move forward with its existing actions for the promotion of inclusion and equal participation in science.

All visitors to the exhibition were able to engage with interactive displays that showcased various resources for multi-sensorial exploration in astronomy from around the world. These include tactile planets, books, and posters (Spain, USA), a multi-sensorial optical telescope (USA), a tactile telescope model (Japan), software that translates astronomical data into audio output (Argentina and Australia), and many more. The inauguration event featured six distinguished invited speakers, who shared their experiences as scientists with impairments and/or disabilities.

With the exhibition, the IAU aims to continue its existing actions for the promotion of inclusion and equal participation in science. "Inspiring Stars" is a collaborative initiative led by the International Astronomical Union and supported by the American

Astronomical Society, along with organisations serving the disabled and/or impaired in national/local communities.

5. Young Astronomers Lunch

The Young Astronomers Lunch was sponsored jointly by the Norwegian Academy of Arts and Sciences and the IAU U.S. National Committee of the National Academy of Sciences on August 23. It provided an opportunity for early career astronomers to network, discuss careers and opportunities, and receive mentorship advice. Ed Guinan presented the opening remarks, followed by discussions at each table.

6. Women in Astronomy Lunch

The International Astronomical Union, through its Executive Committee Working Group on Women in Astronomy, has been a strong advocate for discussing gender and diversity and for supporting initiatives that can lead to a more balanced representation in our community. In this context, the Organizing Committee of the IAU EC WG on Women in Astronomy together with the NOC and LOC of the IAU-GA in Vienna hosted a Women in Astronomy Lunch on August 29, sponsored by the IAU U. S. National Committee of the National Academy of Sciences. It was open to all attendees, and provided an opportunity to network, discuss common issues, and provide advice. The lunch included presentations by Dr. Andreas Keil (European Research Council) and by Prof. Ewine van Dishoeck (IAU President-Elect).

Dr. Ewine van Dishoeck, IAU President-Elect:

Dear colleagues and friends,

It has become a tradition that the president elect of the IAU speaks at the Women in Astronomy lunch, and I am honored to do so. This is the first time in the history of the IAU that for two such lunches in a row, a woman is speaking, and, assuming that Debra Elmegreen will be confirmed as the new President-Elect, that will make it three in 2021. It is clear that over the last decade, the IAU has been making significant progress in terms of gender balance at the top level with the Officers (now 75% women), Executive Committee and Division Presidents (typically 40 50% women). I am also very pleased that about half of the plenary and Invited Discourse speakers are women, and that the new Junior Members enhance diversity at the young end. But, as the statistics show, more work is needed and it requires continuous attention at the highest levels, and this is why the WiA working group reports directly to the Executive Committee.

The first WiA meeting that I attended was at the IAU General Assembly in New Delhi in 1985, in a small dark room. But there I learned in just a few hours a lot about the difficult situations that women from India and elsewhere were facing. For example, they were primarily solar astronomers, simply because women working at night was not an accepted situation by their society at that time. Last year the Executive Committee visited Pune in India and organized a Women in Astronomy afternoon to raise awareness at the institute, university and colleges level: Indian women can now be found in all fields of astronomy, but their fraction is still low compared with other countries according to the 2015 IAU survey. I want to hear more about your stories to get a better overview of where the IAU can help and provide a push in the right direction. I am very pleased that Francesca has taken up this suggestion and organized a separate WiA day this Monday where we heard stories from across the globe, more than we can in just this lunch.

The last three years since the 2015 IAU GA have been a tumultuous and stressful period for our field due to the reports of harassment, bullying and other forms of discrimination. Let me be clear: astronomy worldwide, including the IAU, has to have its house in order, a house where everyone feels welcome and feels safe. All astronomers should abide by our code of conduct, and this should come naturally as part of our humanity rather than having to spell it out word by word. Symbolic compliance by just checking a box at registration is not enough: how you behave is an integral part of who you are. Aretha Franklin's "Respect" song says it all, and is therefore just as powerful today as it was in 1967, and I have fortunately seen a lot of respect at this meeting. At the same time, we should keep an atmosphere in which we can have an open vigorous scientific debate, including sometimes critical questions and disagreements: we need such discussions to advance our science.

I would like to use my energy to move in a positive direction and hear from you what the IAU can do to make further progress for Women in Astronomy. For example, as we wrote in the 2020–2030 Strategic Plan, more actions are needed to retain women in science and make sure that the senior staff complement at universities and institutes reflects that of the younger student/postdoc population in terms of diversity (with diversity being more than just gender balance). This can be done through a top-down approach of setting goals or quotas or special programs for women and several of those have been successful. In my own experience, a bottom-up approach, where "flexibility" is the key word, is equally important: flexibility, for example, to be able to seize unexpected opportunities for a female hire, or by giving women a crucial extra year to build up a better CV and more confidence, to position them better for competitions. It is unfortunate that most of our funding agencies and universities are going exactly the opposite way, by having fewer and fewer grants or budgets that can be used flexibly. The ERC is actually a great exception in this regard.

At newspaper interviews, I often get asked "what advice would you give a young woman just starting in astronomy?" Three responses. First, make sure that you excel in one thing, that you are known as the expert in a certain area, for example Mrs. Youngest galaxies with ALMA. Then you get noticed, much more than being average in lots of things. Second, do whatever you do with a passion, whether inside or outside astronomy. Third, grab opportunities when they arise unexpectedly: I became involved in both ALMA and JWST through unexpected invitations. I could have said no since I had plenty of other things to keep me busy, but I am glad I did say yes, even though each has been a 20–30 year commitment.

Let me end with some more general advice to young people, women and men, on the issues that matter to be successful. I was fortunate to have a PhD advisor and mentor, professor Alexander Dalgarno from Harvard, who was a champion in guiding students and postdocs, and I have been fortunate to guide more than 80 myself. His advice, as summarized by Neal Lane in the *Proceedings of the Dalgarno Celebratory Symposium* book (2009, ed. J. Babb et al., Imperial College Press), was:

- Motivation matters: choose the right problems (this is directly linked to passion)
- Confidence matters: assume nothing is impossible
- Skills matter: use all the tools
- Patience and tenacity matter: never give up
- People matter: human relationships are special
- Integrity matters: stay honest and keep to high standards

To these I would add:

- Fun and family matters: make sure that you enjoy life (inside and outside astronomy) and treasure your family!

Thanks again for your attention, and I look forward to suggestions and activities from the WG on what the IAU can do in the coming three years.

Transactions IAU, Volume XXX
Proc. XXX IAU General Assembly, August 2018
Debra Elmegreen, ed.
doi:10.1017/S1743921323005665

CHAPTER VIII

THIRTIETH GENERAL ASSEMBLY

IAU STATUTES, BYE-LAWS, AND WORKING RULES

Vienna, 30 August 2018

1. IAU Statutes

The IAU Statutes had several changes approved at the August 30 Business Meeting of the XXXth General Assembly. These include the addition of a membership category of Junior Member, described in §11.b, to encourage the engagement of early career astronomers in the activities of the IAU and to help increase diversity. A category of Honorary Member was also added to honor non-astronomers working to further astronomy in their countries, as nominated by National Members. For National Members, the category of "prospective" was changed to "observer." A clause was also added on the possible dissolution of the Union in §25.

I. OBJECTIVE

1. The International Astronomical Union (hereinafter referred to as the Union) is an international nongovernmental organization. Its objective is to promote the science of astronomy in all its aspects.

II. DOMICILE AND INTERNATIONAL RELATIONS

2. The legal domicile of the Union is Paris, France.

3. The Union adheres to, and co-operates with the body of international scientific organizations through the International Council for Science (ICSU). It supports and applies the policies on the Freedom, Responsibility, and Ethics in the Conduct of Science defined by ISC.

III. COMPOSITION OF THE UNION

4. The Union is composed of:

4.a. National Members (adhering organizations)

4.b. Individual Members (adhering persons)

4.c. Junior Members (temporary adhering persons)

4.d. Honorary Members

IV. NATIONAL MEMBERS

5. An organization representing a national professional astronomical community, desiring to promote its participation in international astronomy and supporting the objective of the Union, may adhere to the Union as a National Member.

6. An organization desiring to join the Union as a National Member while developing professional astronomy in the community it represents may do so:

6a. on an interim basis;

6b. on an observer basis.

7. A National Member is admitted to the Union on a permanent, interim, or observer basis by the General Assembly. It may resign from the Union by so informing the General Secretary in writing.

8. A National Member may be either:

8.a. the organization by which scientists of the corresponding nation or territory adhere to ISC or:

8.b. an appropriate National Society or Committee for Astronomy, or:

8.c. an appropriate institution of higher learning.

9. The adherence of a National Member is automatically suspended if its annual contributions, as defined in Articles 23c and 23e below have not been paid for five years; it resumes, upon the approval of the Executive Committee, when the arrears in contributions have been paid in full. After five years of suspension of a National Member, the Executive Committee may recommend to the General Assembly to terminate the Membership. In justified, exceptional cases, the Executive Committee may recommend to the General Assembly to cancel the arrears of a suspended or terminated National Member and resume it.

10. A National Member is admitted to the Union in one of the categories specified in the Bye-Laws.

V. INDIVIDUAL MEMBERS and JUNIOR MEMBERS

11.a. Individual Member:
A professional scientist who is active in some branch of astronomy may be admitted to the Union by the Executive Committee as an Individual Member. An Individual Member may resign from the Union by so informing the General Secretary in writing.

11.b. Junior Member:
A junior scientist who completed the PhD studies and who is in the initial phase of the professional researcher career in some branch of astronomy may be admitted to the Union by the Executive Committee as a Junior Member for a period not exceeding 6 years.

VI. GOVERNANCE

12. The governing bodies of the Union are:

12.a. The General Assembly;

12.b. The Executive Committee; and

12.c. The Officers.

VII. GENERAL ASSEMBLY

13. The General Assembly consists of the National Members and of Individual Members. The General Assembly determines the overall policy of the Union.

13.a. The General Assembly approves the Statutes of the Union, including any changes therein.

13.b.The General Assembly approves Bye-Laws specifying the Rules of Procedure to be used in applying the Statutes.

13.c. The General Assembly elects an Executive Committee to implement its decisions and to direct the affairs of the Union between successive ordinary meetings of the General Assembly. The Executive Committee reports to the General Assembly.

13.d. The General Assembly appoints a standing Finance Committee to advise the Executive Committee on its behalf on budgetary matters between General Assemblies, and to advise the General Assembly on the approval of the budget and accounts of the Union. The Finance Committee consists of not more than 8 members of different national affiliations, including a Chairperson, proposed by the National Members, and remains in office until the end of the next General Assembly.

13.e. The General Assembly appoints a Special Nominating Committee to prepare a suitable slate of candidates for election to the incoming Executive Committee.

13.f. The General Assembly appoints a standing Membership Committee to advise the Executive Committee on its behalf on matters related to the admission of Individual Members. The Membership Committee consists of not more than 8 members of different national affiliations, including a Chairperson, proposed by the National Members, and remains in office until the end of the next General Assembly.

14. Voting at the General Assembly on issues of a primarily scientific nature, as determined by the Executive Committee, is by Individual Members. Voting on all other matters is by National Member. Each National Member authorises a representative to vote on its behalf.

14.a. On questions involving the budget of the Union, the number of votes for each National Member is one greater than the number of its category, referred to in article 10. National Members with interim status, or which have not paid their dues for years preceding that of the General Assembly, may not participate in the voting.

14.b. On questions concerning the administration of the Union, but not involving its budget, each National Member has one vote, under the same condition of payment of dues as in §14.a.

14.c.National Members may vote by correspondence on questions concerning the agenda for the General Assembly.

14.d. A vote is valid only if at least two thirds of the National Members having the right to vote by virtue of article §14.a. participate in it by either casting a vote or signalling an abstention. An abstention. An abstention is not considered a vote cast.

15. The decisions of the General Assembly are taken by an absolute majority of the votes cast. However, a decision to change the Statutes requires the approval of at least two thirds of all National Members having the right to vote by virtue of article §14.a. Where there is an equal division of votes, the President determines the issue.

15.a. To enable the widest possible participation of Individual Members the Executive Committee may decide that voting on certain issues of a primarily scientific nature, as determined by the Executive Committee, shall be open for electronic voting for not more than 31 days counting from the close of the General Assembly at which the issue was raised.

15.b. The Executive Committee shall give Members not less than 3 months notice before the opening of the General Assembly of the intention to open certain issues to electronic voting after the General Assembly.

16. Changes in the Statutes or Bye-Laws can only be considered by the General Assembly if a specific proposal has been duly submitted to the National Members and placed on the Agenda of the General Assembly by the procedure and deadlines specified in the Bye-Laws.

VIII. EXECUTIVE COMMITTEE

17. The Executive Committee consists of the President of the Union, the President-Elect, six Vice-Presidents, the General Secretary, and the Assistant General Secretary, elected by the General Assembly on the proposal of the Special Nominating Committee.

IX. OFFICERS

18. The Officers of the Union are the President, the General Secretary, the President-Elect, and the Assistant General Secretary. The Officers decide short-term policy issues within the general policies of the Union as decided by the General Assembly and interpreted by the Executive Committee.

X. SCIENTIFIC DIVISIONS

19. As an effective means to promote progress in the main areas of astronomy, the scientific work of the Union is structured through its Scientific Divisions. Each Division covers a broad, well-defined area of astronomical science, or deals with international matters of an interdisciplinary nature. As far as practicable, Divisions should include comparable fractions of the Individual and Junior Members of the Union.

20. Divisions are created or terminated by the General Assembly on the recommendation of the Executive Committee. The activities of a Division are organized by an Organizing Committee chaired by a Division President. The Division President and a Vice-President are elected by the General Assembly on the proposal of the Executive Committee, and are ex officio members of the Organizing Committee.

XI. SCIENTIFIC COMMISSIONS

21. Within Divisions, the scientific activities in well-defined disciplines within the subject matter of the Division may be organized through scientific Commissions. In special cases, a Commission may cover a subject common to two or more Divisions and then becomes a Commission of all these Divisions.

22. Commissions are created or terminated by the Executive Committee upon the recommendation of the Organizing Committee(s) of the Division(s) desiring to create or terminate them. The activities of a Commission are organized by an Organizing Committee chaired by a Commission President. The Commission President and a Vice-President are appointed by the Organizing Committee(s) of the corresponding Division(s) upon the proposal of the Organizing Committee of the Commission.

XII. BUDGET AND DUES

23. For each ordinary General Assembly the Executive Committee prepares a budget proposal covering the period to the next ordinary General Assembly, together with the accounts of the Union for the preceding period. It submits these to the Finance Committee for advice before presenting them to the vote of the General Assembly.

23.a.The Finance Committee examines the accounts of the Union from the point of view of responsible expenditure within the intent of the previous General Assembly, as interpreted by the Executive Committee. It also considers whether the proposed budget is adequate to implement the policy of the General Assembly. It submits reports on these matters to the General Assembly before its decisions concerning the approval of the accounts and of the budget.

23.b. The amount of the unit of contribution is decided by the General Assembly as part of the budget approval process.

23.c. Each National Member pays annually a number of units of contribution corresponding to its category. The number of units of contribution for each category shall be specified in the Bye-Laws.

23.d. A vote on matters under article 23 is valid only if at least two thirds of the National Members having the right to vote by virtue of article §14.a. cast a vote. In all cases an abstention is not a vote, but a declaration that the Member declines to vote.

23.e. National Members having interim status pay annually one half unit of contribution.

23.f. National Members having observer status pay no contribution.

23.g. The payment of contributions is the responsibility of the National Members. The liability of each National Members in respect of the Union is limited to the amount of contributions due through the current year.

XIII. EMERGENCY POWERS

24. If, through events outside the control of the Union, circumstances arise in which it is impracticable to comply fully with the provisions of the Statutes and Bye-Laws of the Union, the Executive Committee and Officers, in the order specified below, shall take such actions as they deem necessary for the continued operation of the Union. Such action shall be reported to all National Members as soon as this becomes practicable, until an ordinary or extraordinary General Assembly can be convened. The following is the order of authority: The Executive Committee in meeting or by correspondence; the President of the Union; the General Secretary; or failing the practicability or availability of any of the above, one of the Vice-Presidents.

XIV. DISSOLUTION OF THE UNION

25. A decision to dissolve the Union is only valid if taken by the General Assembly with the approval of three quarters of the National Members having the right to vote by virtue of article §14.a. Such a decision shall specify a procedure for settling any debts and disposing of any assets of the Union. Upon dissolution, all assets will be used exclusively for charitable purposes, including but not limited to support to colleagues from economically less privileged countries, providing education and outreach activities to young scientists and for education and outreach activities at large.

XV. FINAL CLAUSES

26. These Statutes enter into force on 21 August 2012 and revised 30 August 2018.

27. The present Statutes are published in French and English versions. For legal purposes, the French version is authoritative.

2. Statuts de l'UAI, Vienna, 30 Août 2018

I. OBJECTIF

1. L'Union astronomique internationale (dénommée ci-après ≪ l'Union ≫) est une organisation non gouvernementale internationale. Son objectif est de promouvoir la science de l'astronomie sous tous ses aspects.

II. DOMICILIATION ET RELATIONS INTERNATIONALES

2. Le domicile légal de l'Union est situé à Paris, en France.

3. L'Union adhère à, et coopère avec, l'ensemble des organisations scientifiques internationales à travers le Conseil international pour la science (ISC). Elle soutient et applique les directives définies par ISC en matière de liberté, de responsabilité et d'éthique dans les activités scientifiques.

III. COMPOSITION DE L'UNION

4. L'Union se compose de:

4.a. Membres Nationaux (organisations)

4.b. Membres individuels (personnes physiques)

4.c. Membres juniors (personnes adhérant temporairement)

4.b. Membres honoraires

IV. MEMBRES NATIONAUX

5. Une organisation représentant une communauté astronomique professionnelle nationale, désireuse de développer sa participation sur la scène de l'astronomie internationale et partageant l'objectif de l'Union, peut adhérer à l'Union en qualité de Membre national.

6. Une organisation désireuse de rejoindre l'Union en qualité de Membre national tout en développant l'astronomie professionnelle dans la communauté qu'elle représente peut le faire :

6.a. de manière temporaire ;

6.b. en tant qu'observateur.

7. Un Membre national est admis au sein de l'Union de manière permanente, temporaire ou en tant qu'observateur par l'Assemblée générale. Il peut se retirer de l'Union en informant le Secrétaire Général de son retrait par écrit.

8. Un Membre National peut être :

8.a. l'organisation par laquelle les scientifiques du pays ou du territoire correspondant adhérent à ISC;

8.b. une Société ou un Comité national(e) compétent(e) en matière d'astronomie:

8.c. un établissement d'enseignement supérieur compétent.

9. L'adhésion d'un Membre national est automatiquement suspendue si ses cotisations annuelles, telles que définies aux points 23c et 23e ci-dessous n'ont pas été payées pendant cinq ans ; elle sera rétablie, sur approbation du Comité exécutif, lorsque les arriérés relatifs à ses cotisations auront été payés en totalité. Après cinq ans de suspension d'un Membre national, le Comité exécutif peut recommander à l'Assemblée générale l'exclusion de ce Membre. Si cela est justifié, dans des cas exceptionnels, le Comité exécutif peut recommander à l'Assemblée générale d'annuler les arriérés d'un Membre national dont l'adhésion est suspendue ou a pris fin et de la rétablir.

10. Un Membre national est admis au sein de l'Union au titre de l'une des catégories précisées dans le Règlement intérieur.

V. MEMBRES INDIVIDUELS ET MEMBRES JUNIORS

11.a. Membre individuel :

Un scientifique professionnel exerçant son activité dans un des domaines de l'astronomie peut être admis au sein de l'Union par le Comité exécutif en qualité de Membre individuel. Un Membre individuel peut quitter l'Union en informant par écrit le Secrétaire général de son retrait.

11.b. Membre junior :

Un scientifique junior qui a achevé un doctorat et se trouve dans la phase initiale de sa carrière de chercheur professionnel dans un des domaines de l'astronomie peut être admis au sein de l'Union par le Comité exécutif en qualité de Membre junior pour une durée ne dépassant pas 6 ans.

VI. GOUVERNANCE

12. Les organes de gouvernance de l'Union sont les suivants :

12.a. l'Assemblée générale;

12.b. le Comité exécutif;

12.c. les Membres du Bureau du Comité exécutif.

VII. ASSEMBLÉE GÉNÉRALE

13. L'Assemblée générale se compose des Membres nationaux et des Membres individuels et junior. Elle détermine les grandes orientations de l'Union.

13.L'Assemblée générale approuve les Statuts de l'Union, y compris tout changement apporté à ceux-ci.

13.b. L'Assemblée générale approuve le Règlement intérieur qui précise les Règles de procédure devant être suivies lors de l'application des Statuts.

13.c. L'Assemblée générale élit un Comité exécutif chargé de mettre en œuvre ses décisions et de mener les affaires de l'Union d'une Assemblée générale à l'autre. Le Comité exécutif fait rapport à l'Assemblée générale.

13.d. L'Assemblée générale instaure un Comité des finances pour conseiller le Comité exécutif en son nom sur les questions budgétaires entre les Assemblées générales et pour proposer ses recommandations à l'Assemblée générale concernant l'approbation du budget et des comptes de l'Union. Le Comité des finances se compose au plus de huit membres de différentes représentations nationales, dont un Président, proposés par les Membres nationaux, et il reste en place jusqu'à la fin de l'Assemblée générale suivante.

13.e. L'Assemblée générale instaure un Comité spécial des nominations chargé de préparer une liste de candidats en vue de l'élection du Comité exécutif suivant.

13.f. L'Assemblée générale instaure un Comité d'admission pour permettre au Comité exécutif de se prononcer en son nom sur l'admission de nouveaux Membres individuels et junior. Le Comité d'admission se compose au plus de huit membres de différentes représentations nationales, dont un Président, proposés par les Membres nationaux, et il reste en place jusqu'à la fin de l'Assemblée générale suivante.

14. Lors de l'Assemblée générale, les Membres individuels et les Membres juniors votent sur les questions de nature essentiellement scientifique, telles que déterminées par le Comité exécutif. Les Membres nationaux votent sur toutes les autres questions. Chaque Membre national nomme un représentant pour voter en son nom.

14.a. Concernant les questions ayant trait au budget de l'Union, le nombre de voix pour chaque Membre national est supérieur d'une unité au nombre définissant sa catégorie, telle que visée au point 10. Les Membres nationaux ayant un statut temporaire ou prospectif, ou qui ne sont pas à jour de leur cotisation au moment de l'Assemblée générale, ne peuvent pas prendre part au vote.

14.b. Concernant les questions relatives à l'administration de l'Union, mais n'ayant pas trait au budget, chaque Membre national dispose d'une voix, selon les mêmes conditions que celles relatives au paiement des cotisations mentionnées au point 14.a.

14.c. Les Membres nationaux peuvent voter par correspondance sur des questions relatives à l'ordre du jour de l'Assemblée générale.

14.d. Un vote n'est valable que si au moins deux tiers des Membres nationaux ayant le droit de voter en vertu du point 14.a. y participent, soit en votant, soit en se prononçant par une abstention. Une abstention n'est pas considérée comme l'expression d'un vote.

15. Les décisions de l'Assemblée générale sont prises à la majorité absolue des voix. Cependant, la décision de modifier les Statuts nécessite l'approbation d'au moins les deux tiers de tous les Membres nationaux ayant le droit de voter en vertu du point 14.a. Lorsqu'il y a égalité des voix, il revient au Président de statuer sur l'issue du vote.

15.a. Afin de favoriser la plus large participation possible des Membres individuels et junior, le Comité exécutif pourra décider que le vote portant sur certaines questions de nature essentiellement scientifique, telles que déterminées par le Comité exécutif, pourront faire l'objet d'un vote électronique 31 jours au plus à partir de la clôture de l'Assemblée générale à laquelle la question a été soulevée.

15.b. Le Comité exécutif signalera aux membres au moins trois mois avant l'ouverture de l'Assemblée générale les questions devant être soumises à un vote électronique après l'Assemblée générale.

16. Les changements apportés aux Statuts ou au Règlement intérieur ne peuvent être examinés par l'Assemblée générale que si une proposition spécifique a été dûment soumise aux Membres nationaux et inscrite à l'ordre du jour de l'Assemblée générale selon la procédure et dans les délais précisés dans le Règlement intérieur.

VIII. COMITE EXÉCUTIF

17. Le Comité exécutif se compose du Président de l'Union, du Président désigné, de six Vice-Présidents, du Secrétaire général et du Secrétaire général adjoint, élus par l'Assemblée générale sur proposition du Comité spécial des nominations.

IX. MEMBRES DU BUREAU

18. Les Membres du Bureau de l'Union sont le Président, le Secrétaire général, le Président désigné et le Secrétaire général adjoint. Les Membres du Bureau prennent des décisions sur le court terme dans le cadre des orientations générales de l'Union telles que décidées par l'Assemblée générale et interprétées par le Comité exécutif.

X. DIVISIONS SCIENTIFIQUES

19. Pour favoriser efficacement les progrès dans les principaux domaines de l'astronomie, le travail scientifique de l'Union est réparti entre ses Divisions scientifiques. Chaque Division couvre un large domaine bien défini de l'astronomie, ou traite de sujets internationaux de nature interdisciplinaire. Autant que possible, les Divisions comprennent chacune un nombre comparable de Membres individuels et junior de l'Union.

20. Les Divisions sont créées ou supprimées par l'Assemblée générale sur recommandation du Comité exécutif. Les activités d'une Division sont organisées par un Comité de pilotage présidé par le Président de Division. Le Président de Division et un Vice-Président sont élus par l'Assemblée générale sur proposition du Comité exécutif, et sont membres d'office du Comité de pilotage.

XI. COMMISSIONS SCIENTIFIQUES

21. Au sein des Divisions, les activités scientifiques réparties en disciplines bien définies peuvent être organisées dans le cadre de Commissions scientifiques. Dans certains cas, une Commission peut couvrir une discipline commune à deux ou plusieurs Divisions et devient alors une Commission de l'ensemble de ces Divisions.

22. Les Commissions sont créées ou supprimées par le Comité exécutif sur recommandation du ou des Comité(s) de pilotage de la ou des Division(s) désirant les créer ou les supprimer. Les activités d'une Commission sont organisées par un Comité d'organisation présidé par le Président de Commission. Le Président de Commission et un Vice-Président sont nommés par le ou les Comité(s) de pilotage de la ou des Division(s) correspondante(s) sur proposition du Comité d'organisation de la Commission.

XII. BUDGET ET COTISATIONS

23. En vue de chaque Assemblée générale ordinaire, le Comité exécutif prépare une proposition de budget couvrant la période jusqu'à l'Assemblée générale ordinaire suivante, accompagnée des comptes de l'Union pour la période précédente. Il les soumet au Comité des finances pour avis avant leur soumission au vote de l'Assemblée générale.

23.a. Le Comité des finances examine les comptes de l'Union au regard des dépenses approuvées par l'Assemblée générale précédente, selon l'interprétation faite par le Comité exécutif. Il se penche aussi sur l'adéquation ou non du budget proposé en vue de la mise en oeuvre des orientations décidées par l'Assemblée générale. Il soumet son rapport à l'Assemblée générale avant ses décisions concernant l'approbation des comptes et du budget.

23.b. Le montant de l'unité de cotisation est décidé par l'Assemblée générale au titre du processus d'approbation du budget.

23.c. Chaque Membre national verse annuellement un nombre d'unités de cotisation correspondant à sa catégorie. Le nombre d'unités de cotisation pour chaque catégorie est précisé dans le Règlement intérieur.

23.d. Un vote sur les questions relevant du point 23 n'est valable que si les deux tiers des représentants des Membres nationaux ayant le droit de vote en vertu du point 14.a. prennent part à ce vote. Dans tous les cas, une abstention ne constitue pas un vote, mais une déclaration selon laquelle le Membre national décide de ne pas voter.

23.e. Les Membres nationaux ayant un statut temporaire payent annuellement une demi-unité de cotisation.

23.f. Les Membres nationaux ayant un statut d'observateur ne payent aucune cotisation.

23.g. Le paiement des cotisations relève de la responsabilité des Membres nationaux. La responsabilité de chaque Membre national relative à l'Union se limite au montant des cotisations dues au cours de l'année.

XIII. POUVOIRS SPÉCIAUX

24. Si, en cas d'événements hors du contrôle de l'Union, les circonstances font qu'il est impossible de respecter pleinement les dispositions des Statuts et du Règlement intérieur de l'Union, le Comité exécutif et les Membres du Bureau, dans l'ordre précisé ci-après, agiront comme ils l'estimeront nécessaire pour assurer la continuité du fonctionnement de l'Union. De telles décisions seront rapportées à l'ensemble des Membres nationaux dès que cela sera possible, jusqu'à ce qu'une Assemblée générale ordinaire ou extraordinaire soit convoquée.

L'ordre hiérarchique est le suivant : le Comité exécutif, en réunion ou par correspondance; le Président de l'Union; le Secrétaire général; ou à défaut de possibilité ou de disponibilité des entités précédentes, un des Vice-Présidents.

XIV. DISSOLUTION DE L'UNION

25. La décision de dissoudre l'Union n'est recevable que si elle est prise par l'Assemblée générale avec l'approbation des trois quarts des Membres nationaux ayant le droit de voter en vertu du point 14.a. Une telle décision précisera une procédure visant à apurer les comptes et à disposer des actifs de l'Union. Après dissolution, tous les actifs seront

utilisés exclusivement à des fins caritatives, y compris, sans que cela soit limitatif, pour apporter un soutien à des collègues de pays moins privilégiés d'un point de vue économique, en fournissant des activités d'enseignement ou de sensibilisation à de jeunes scientifiques ou au grand public.

XV. CLAUSES FINALES

26. Les présents Statuts entrent en vigueur le 21 Août 2012 et révisé le 30 Août 2018.

27. Les présents Statuts sont publiés en français et en anglais. À toutes fins juridiques, la version française fait autorité.

3. IAU Bye-Laws

I. MEMBERSHIP

1. An application for admission to the Union as a National Member shall be submitted to the General Secretary by the proposing organization at least eight months before the next ordinary General Assembly.

2. The Executive Committee shall examine the application and resolve any outstanding issues concerning the nature of the proposed National Member and the category of membership (§VII.25). Subsequently, the Executive Committee shall forward the application to the General Assembly for decision, with its recommendation as to its approval or rejection.

3. The Executive Committee shall examine any proposal by a National Member to change its category of adherence to a more appropriate level. If the Executive Committee is unable to approve the request, either party may refer the matter to the next General Assembly.

4. Individual Members, Junior Members and Honorary Members are admitted by the Executive Committee upon the nomination of a National Member or the President of a Division. The Executive Committee shall publish the criteria and procedures for membership, and shall consult the Membership Committee before admitting new Individual Members.

II. GENERAL ASSEMBLY

5. The ordinary General Assembly meets, as a rule, once every three years. Unless determined by the previous General Assembly, the place and date of the ordinary General Assembly shall be fixed by the Executive Committee and be communicated to the National Members at least one year in advance.

6. The President may summon an extraordinary General Assembly with the consent of the Executive Committee, and must do so at the request of at least one third of the National Members. The date, place, and agenda of business of an extraordinary General Assembly must be communicated to all National Members at least two months before the first day of the Assembly.

7. Matters to be decided upon by the General Assembly shall be submitted for consideration by those concerned as follows, counting from the first day of the General Assembly:

7.a. A motion to amend the Statutes or Bye-Laws may be submitted by a National Member or by the Executive Committee. Any such motion shall be submitted to the General Secretary at least nine months in advance and be forwarded, with the recommendation of the Executive Committee as to its adoption or rejection, to the National Members at least six months in advance.

7.b. The General Secretary shall distribute the draft budget prepared by the Executive Committee to the National Members at least eight months in advance. Any motion to modify this budget, or any other matters pertaining to it, shall be submitted to the General Secretary at least six months in advance. The Executive Committee shall consider whether or not to adopt any such motion in a modified budget, which shall be distributed to the National Members at least four months in advance. Should the Executive Committee decide to reject the motion it shall also be submitted to the General Assembly with the reasons for its rejection.

7.c. Any motion or proposal concerning the administration of the Union, and not affecting the budget, by a National Member, or by the Organizing Committee of a Scientific Division of the Union, shall be placed on the Agenda of the General Assembly, provided it is submitted to the General Secretary, in specific terms, at least six months in advance.

7.d. Any motion of a scientific character submitted by a National Member, a Scientific Division of the Union, or by an ICSU Scientific Committee or Program on which the Union is formally represented, shall be placed on the Agenda of the General Assembly, provided it is submitted to the General Secretary, in specific terms, at least six months in advance.

7.e. The complete agenda, including all such motions or proposals, shall be prepared by the Executive Committee and submitted to the National Members at least four months in advance.

8. The President may invite representatives of other organizations, scientists in related fields, and young astronomers to participate in the General Assembly. Subject to the agreement of the Executive Committee, the President may authorise the General Secretary to invite representatives of other organizations, and the National Members or other appropriate IAU bodies to invite scientists in related fields and young astronomers.

III. SPECIAL NOMINATING COMMITTEE

9. The Special Nominating Committee consists of the President and past President of the Union, a member proposed by the retiring Executive Committee, and four members selected by the representatives of the National Members from up to twelve candidates proposed by Presidents of Divisions, with due regard to an appropriate distribution over the major branches of astronomy.

9.a. Except for the President and immediate past President, present and former members of the Executive Committee shall not serve on the Special Nominating Committee. No two members of the Special Nominating Committee shall belong to the same nation or National Member.

9.b. The General Secretary and the Assistant General Secretary participate in the work of the Special Nominating Committee in an advisory capacity, and the President-Elect may participate as an observer.

10. The Special Nominating Committee is appointed by the General Assembly, to which it reports directly. It assumes its duties immediately after the end of the General Assembly and remains in office until the end of the ordinary General Assembly next following that of its appointment, and it may fill any vacancy occurring among its members.

IV. OFFICERS AND EXECUTIVE COMMITTEE

11. Terms of office:

11.a. The President of the Union remains in office until the end of the ordinary General Assembly next following that of election. The President-Elect succeeds the President at that moment.

11.b. The General Secretary and the Assistant General Secretary remain in office until the end of the ordinary General Assembly next following that of their election. Normally the Assistant General Secretary succeeds the General Secretary, but both officers may be re-elected for another term.

11.c. The Vice-Presidents remain in office until the end of the ordinary General Assembly following that of their election. They may be immediately re-elected once to the same office.

11.d. The elections take place at the last session of the General Assembly, the names of the candidates proposed having been announced at a previous session.

12. The Executive Committee may fill any vacancy occurring among its members. Any person so appointed remains in office until the end of the next ordinary General Assembly.

13. The past President and General Secretary become advisers to the Executive Committee until the end of the next ordinary General Assembly. They participate in the work of the Executive Committee and attend its meetings without voting rights.

14. The Executive Committee shall formulate Working Rules to clarify the application of the Statutes and Bye-Laws. Such Working Rules shall include the criteria and procedures by which the Executive Committee will review applications for Individual Membership; standard Terms of Reference for the Scientific Commissions of the Union; rules for the administration of the Union's financial affairs by the General Secretary; and procedures by which the Executive Committee may conduct business by electronic or other means of correspondence. The Working Rules shall be published electronically and in the Transactions of the Union.

15. The Executive Committee appoints the Union's official representatives to other scientific organizations.

16. The Officers and members of the Executive Committee cannot be held individually or personally liable for any legal claims or charges that might be brought against the Union.

V. SCIENTIFIC DIVISIONS

17. The Divisions of the Union shall pursue the scientific objects of the Union within their respective fields of astronomy. Activities by which they do so include the encouragement and organization of collective investigations, and the discussion of questions relating to international agreements, cooperation, or standardization. They shall report to each General Assembly on the work they have accomplished and such new initiatives as they are undertaking.

18. Each Scientific Division shall consist of:

18.a. An Organizing Committee, normally of 6-12 persons, including the Division President and Vice-President, and a Division Secretary appointed by the Organizing Committee from among its members. The Committee is responsible for conducting the business of the Division.

18.b. Members of the Union accepted by the Organizing Committee in recognition of their special experience and interests.

19. Normally, the Division President is succeeded by the Vice-President at the end of the General Assembly following their election, but both may be re-elected for a second term. Before each General Assembly, the Organizing Committee shall organize an election from among the membership, by electronic or other means suited to its scientific structure, of a new Organizing Committee to take office for the following term. Election procedures should, as far as possible, be similar among the Divisions and require the approval of the Executive Committee.

20. Each Scientific Division may structure its scientific activities by creating a number of Commissions. In order to monitor and further the progress of its field of astronomy, the Division shall consider, before each General Assembly, whether its Commission structure serves its purpose in an optimum manner. It shall subsequently present its proposals for the creation, continuation or discontinuation of Commissions to the Executive Committee for approval.

21. With the approval of the Executive Committee, a Division may establish Working Groups to study well-defined scientific issues and report to the Division. Unless specifically re-established by the same procedure, such Working Groups cease to exist at the following General Assembly.

VI. SCIENTIFIC COMMISSIONS

22. A Scientific Commission shall consist of:

22.a. A President and an Organizing Committee consisting of 4-8 persons elected by the Commission membership, subject to the approval of the Organizing Committee of the Division. Additionally a Commission Secretary should be appointed by the Organizing Committee from among its members;

22.b. Members of the Union, accepted by the Organizing Committee, in recognition of their special experience and interests.

23. A Commission is initially created for a period of six years. The parent Division may recommend its continuation for additional periods of three years at a time, if sufficient justification for its continued activity is presented to the Division and the Executive Committee. The activities of a Commission are governed by Terms of Reference, which are based on a standard model published by the Executive Committee and are approved by the Division.

24. With the approval of the Division, a Commission may establish Working Groups to study well-defined scientific issues and report to the Commission. Unless specifically re-appointed by the same procedure, such Working Groups cease to exist at the following General Assembly.

VII. ADMINISTRATION AND FINANCES

25. Each National Member pays annually to the Union a number of units of contribution corresponding to its category as specified below. National Members with interim status pay annually one half unit of contribution, and those with prospective status pay no dues.

Categories as defined in Statutes, §10:	I	II	III	IV	V	VI	VII	VIII	IX	X	XI	XII
Number of units of contribution:	1	2	4	6	10	14	20	27	35	45	60	80

26. The income of the Union is to be devoted to its objects, including:

26.a. the promotion of scientific initiatives requiring international co-operation;

26.b. the promotion of the education and development of astronomy world-wide;

26.c. the costs of the publications and administration of the Union.

27. Funds derived from donations are reserved for use in accordance with the instructions of the donor(s). Such donations and associated conditions require the approval of the Executive Committee.

28. The General Secretary is the legal representative of the Union. The General Secretary is responsible to the Executive Committee for not incurring expenditure in excess of the amount specified in the budget as approved by the General Assembly.

29. The General Secretary shall consult with the Finance Committee (cf. Statutes §13.d.) in preparing the accounts and budget proposals of the Union, and on any other matters of major importance for the financial health of the Union. The comments and advice of the Finance Committee shall be made available to the Officers and Executive Committee as specified in the Working Rules.

30. An Administrative office, under the direction of the General Secretary, conducts the correspondence, administers the funds, and preserves the archives of the Union.

31. The Union has copyright to all materials printed in its publications, unless otherwise arranged.

VIII. FINAL CLAUSES

32. These Bye-Laws enter into force on 21 August 2012.

33. The present Bye-Laws are published in French and English versions. For legal purposes, the French version is authoritative.

4. Règlement intérieur de l'UAI, Vienna, 30 Août 2018

I. ADHÉSION

1. Une demande d'adhésion à l'Union en qualité de Membre national doit être soumise au Secrétaire général par l'organisation demandeuse au moins huit mois avant l'Assemblée générale ordinaire suivante..

2. Le Comité exécutif examine la demande et résout toute question en suspens concernant la nature du Membre national proposé et la catégorie d'adhésion (§VII.25). Il transmet la demande à l'Assemblée générale afin que celle-ci prenne sa décision, en l'accompagnant de sa recommandation quant à son approbation ou son rejet.

3. Le Comité exécutif examine chaque proposition de Membre national et l'invite éventuellement à changer sa catégorie d'adhésion à un niveau plus approprié. Si le Comité exécutif ne peut approuver la demande, l'une ou l'autre des parties peut soumettre la question à l'Assemblée générale suivante.

4. Les Membres individuels, les Membres juniors et les Membres honoraires sont admis par le Comité exécutif sur proposition par un Membre national ou le Président d'une Division. Le Comité exécutif publie les critères et procédures d'adhésion, et consulte le Comité d'admission avant d'admettre de nouveaux membres.

II. ASSEMBLÉE GÉNERALE

5. L'Assemblée générale ordinaire se réunit, en règle générale, une fois tous les trois ans. Si le lieu et la date de l'Assemblée générale ordinaire n'ont pas été fixés par l'Assemblée générale précédente, ils le sont par le Comité exécutif et sont communiqués aux Membres nationaux au minimum un an à l'avance.

6. Le Président peut convoquer une Assemblée générale extraordinaire avec l'autorisation du Comité exécutif, et doit le faire à la demande d'au moins un tiers des Membres nationaux. La date, le lieu et l'ordre du jour d'une Assemblée générale extraordinaire doivent être communiqués à tous les Membres nationaux au minimum deux mois avant le premier jour de l'Assemblée.

7. Les points devant être examinés par l'Assemblée générale sont transmis par les personnes concernées selon les modalités ci-après, le délai s'entendant à partir du premier jour de l'Assemblée générale:

7.a. une motion visant à modifier les Statuts ou le Règlement intérieur peut être soumise par un Membre national ou par le Comité exécutif. Elle doit être adressée au Secrétaire général au minimum neuf mois à l'avance et transmise, avec la recommandation du Comité exécutif quant à son adoption ou son rejet, aux Membres nationaux au minimum six mois à l'avance.

7.b. Le Secrétaire Général diffuse le budget prévisionnel préparé par le Comité exécutif aux Membres nationaux à-peu-près six mois à l'avance (en générale début de Mars). Toute motion visant à modifier ce budget, ou toute autre question y afférente, doit être soumise au Secrétaire général au minimum quatre mois à l'avance (en générale mi-Avril). Le Comité exécutif décide de modifier ou non le budget en conséquence, l'éventuelle version révisée devant être diffusée aux Membres nationaux à-peu-près quatre mois à l'avance (en générale fin Avril après le meeting du Comité exécutif). Si le Comité exécutif décide de rejeter la motion, celle-ci doit néanmoins être soumise à l'Assemblée générale accompagnée des motifs de son rejet.

7.c. Toute motion ou proposition concernant l'administration de l'Union, et n'ayant pas d'incidence sur le budget, présentée par un Membre national ou par le Comité d'organisation d'une Division scientifique de l'Union, est inscrite à l'ordre du jour de l'Assemblée générale, à condition qu'elle soit soumise au Secrétaire général, en des termes spécifiques, au minimum six mois à l'avance.

7.d. Toute motion à caractère scientifique soumise par un Membre national, par une Division scientifique de l'Union, ou par un Programme ou un Comité scientifique de ISC au sein duquel l'Union est représentée officiellement, est inscrite à l'ordre du jour de l'Assemblée générale, à condition qu'elle soit soumise au Secrétaire général, en des termes spécifiques, au minimum six mois à l'avance.

7.e. L'ordre du jour complet, incluant toutes ces motions ou propositions, est préparé par le Comité exécutif et soumis aux Membres nationaux au minimum quatre mois à l'avance.

8. Le Président peut inviter des représentants d'autres organisations, des scientifiques actifs dans des domaines connexes et de jeunes astronomes à participer à l'Assemblée générale. Avec l'accord du Comité exécutif, le Président peut autoriser le Secrétaire général à inviter des représentants d'autres organisations, et les Membres nationaux ou autres organes de l'UAI appropriés à inviter des scientifiques actifs dans des domaines connexes et de jeunes astronomes.

III. COMITÉ SPÉCIAL DES NOMINATIONS

9. Le Comité spécial des nominations est constitué du Président et de l'ancien Président de l'Union, d'un membre proposé par le Comité exécutif sortant, et de quatre membres sélectionnés par les représentants des Membres nationaux parmi les candidats (jusqu'à douze) proposés par les Présidents de Division, en tenant dûment compte d'une répartition appropriée dans les principaux domaines de l'astronomie.

9.a. À l'exception du Président et de l'ancien Président immédiat, les membres actuels et anciens du Comité exécutif ne peuvent siéger dans le Comité spécial des nominations. Deux membres du Comité spécial des nominations ne peuvent provenir du même pays ou du même Membre national.

9.b. Le Secrétaire général et le Secrétaire général adjoint participent aux travaux du Comité spécial des nominations en qualité de conseillers, et le Président désigné peut participer en qualité d'observateur.

10. Le Comité spécial des nominations est instauré par l'Assemblée générale dont il dépend directement. Il démarre ses activités immédiatement après la fin de l'Assemblée générale et les mène jusqu'à la fin de l'Assemblée générale ordinaire suivant celle de sa nomination, et il peut combler toute vacance se produisant parmi ses membres.

IV. MEMBRES DU BUREAUET COMITÉ EXÉCUTIF

11. Mandat:

11.a. Le Président de l'Union demeure en poste jusqu'à la fin de l'Assemblée générale ordinaire suivant celle de son élection, puis le Président désigné lui succède..

11.b. Le Secrétaire général et le Secrétaire général adjoint demeurent en poste jusqu'à la fin de l'Assemblée générale ordinaire suivant celle de leur élection. Normalement, le Secrétaire général adjoint succède au Secrétaire général, mais ces deux membres du Bureau peuvent être réélus pour un autre mandat.

11.c. Les Vice-Présidents demeurent en poste jusqu'à la fin de l'Assemblée générale ordinaire suivant celle de leur élection. Ils peuvent être immédiatement réélus une fois à la même fonction.

11.d. Les élections ont lieu au cours de la dernière session de l'Assemblée générale, les noms des candidats proposés ayant été annoncés au cours d'une session préalable.

12. Le Comité exécutif peut combler toute vacance se produisant parmi ses membres. Toute personne ainsi nommée demeure en poste jusqu'à la fin de l'Assemblée générale ordinaire suivante.

13. Les anciens Président et Secrétaire général deviennent conseillers du Comité exécutif jusqu'à la fin de l'Assemblée générale ordinaire suivante. Ils participent aux activités du Comité exécutif et assistent à ses réunions sans droits de vote.

14. Le Comité exécutif établit des Règles de travail en vue de préciser l'application des Statuts et du Règlement intérieur. Ces Règles de travail doivent inclure les critères et procédures selon lesquels le Comité exécutif étudie les demandes d'adhésion en tant que Membre individuel ou junior; le mandat standard des Commissions scientifiques de l'Union; les règles d'administration des affaires financières de l'Union par le Secrétaire général et les procédures selon lesquelles le Comité exécutif peut mener ses activités par des moyens de correspondance électroniques ou autres. Les Règles de travail font l'objet d'une diffusion électronique et sont publiées dans les Transactions de l'Union.

15. Le Comité exécutif nomme les représentants officiels de l'Union auprès d'autres organisations scientifiques..

16. Les Membres du Bureau et du Comité exécutif ne peuvent être tenus individuellement ou personnellement responsables pour toute réclamation ou toutes poursuites qui seraient engagées à l'encontre de l'Union.

V. DIVISIONS SCIENTIFIQUES

17. Les Divisions de l'Union poursuivent les objectifs scientifiques de l'Union dans leurs domaines d'astronomie respectifs. Les activités correspondantes incluent l'encouragement et l'organisation de travaux de recherche collectifs, et la discussion de questions en rapport avec les accords internationaux, la coopération ou la normalisation. Les Divisions rendent compte à chaque Assemblée générale des activités qu'elles ont menées et des initiatives qu'elles ont prises.

18. Chaque Division scientifique comprend:

18.a. un Comité de pilotage, constitué normalement de 6 à 12 personnes, dont le Président et le Vice-Président de la Division, et un Secrétaire de Division nommé par le Comité de pilotage parmi ses membres. Le Comité de pilotage est responsable de la conduite des activités de la Division.

18.b. les membres de l'Union acceptés par le Comité de pilotage en reconnaissance de leurs expériences et intérêts particuliers.

19. Normalement, le Vice-Président succède au Président de la Division à la fin de l'Assemblée générale suivant leur élection, mais tous les deux peuvent être réélus pour un second mandat. Avant chaque Assemble générale, le Comité de pilotage organise une élection parmi les membres, par des moyens électroniques ou autres adaptés à sa structure scientifique, d'un nouveau Comité de pilotage devant prendre ses fonctions pour le mandat suivant. Les procédures d'élection doivent, dans la mesure du possible, être similaires parmi les Divisions et sont approuvées par le Comité exécutif.

20. Chaque Division scientifique peut structurer ses activités scientifiques en créant plusieurs Commissions. Afin de suivre l'évolution de son domaine d'astronomie, la Division doit déterminer, avant chaque Assemblée générale, si sa structure en Commissions sert au mieux son objectif. Elle doit ensuite présenter au Comité exécutif, pour approbation, ses propositions de création, maintien ou suppression de Commissions.

21. Avec l'approbation du Comité exécutif, une Division peut constituer des Groupes de travail chargés d'étudier des questions scientifiques bien définies et d'en rendre compte à la Division. Sauf s'ils sont spécifiquement réinstaurés par la même procédure, ces Groupes de travail cesseront d'exister lors de l'Assemblée générale qui suit immédiatement.

VI. COMMISSIONS SCIENTIFIQUES

22. Une Commission scientifique comprend:

22.a. un Président, un Vice-Président et un Comité d'organisation constitué de 4 à 8 personnes élues par les membres de la Commission, sous réserve de l'approbation du Comité de pilotage de la Division. De plus, un Secrétaire de Commission sera nommé par le Comité d'organisation parmi ses membres;

22.b. les membres de l'Union acceptés par le Comité d'organisation en reconnaissance de leurs expériences et intérêts particuliers.

23. Une Commission est initialement créée pour une période de six ans. La Division mère peut recommander son maintien durant des périodes additionnelles de trois ans, si une justification suffisante de son maintien est présentée à la Division et au Comité exécutif. Les activités d'une Commission sont régies par son mandat, lequel est basé sur un modèle standard publié par le Comité exécutif et approuvé par la Division..

24. Avec l'approbation de la Division, une Commission peut constituer des Groupes de travail chargés d'étudier des questions scientifiques bien définies et d'en rendre compte à la Commission. Sauf s'ils sont spécifiquement réinstaurés par la même procédure, ces Groupes de travail cesseront d'exister lors de l'Assemblée générale qui suit immédiatement.

VII. ADMINISTRATION ET FINANCES

25. Chaque Membre national paie chaque année à l'Union un nombre d'unités de cotisation correspondant à sa catégorie telle que spécifiée ci-après. Les Membres nationaux ayant un statut temporaire paient chaque année une demi-unité de cotisation, et ceux ayant un statut d'observateur n'en paient aucune.

Catégories telles que mentionnées au point 10 des Statuts, §10:

I	II	III	IV	V	VI	VII	VIII	IX	X	XI	XII
1	2	4	6	10	14	20	27	35	45	60	80

Nombre d'unités de cotisation

26. Les revenus de l'Union doivent être utilisés pour atteindre ses objectifs, notamment:

26.a. la promotion d'initiatives scientifiques nécessitant une coopération internationale;

26.b.la promotion de l'enseignement de l'astronomie et de son développement dans le monde entier;

26.c. la couverture des coûts des publications et d'administration de l'Union.

27. Les fonds provenant de donations doivent être strictement utilisés conformément aux instructions du ou des donateur(s). De telles donations et les conditions associées sont approuvées par le Comité exécutif.

28. Le Secrétaire général est le représentant légal de l'Union. Il est responsable vis-à-vis du Comité exécutif de ne pas engager de dépenses supérieures au montant spécifié dans le budget tel qu'approuvé par l'Assemblée générale.

29. Le Secrétaire général doit consulter le Comité des finances (cf. Statuts, §13.d) lors de la préparation des comptes et des propositions de budget de l'Union, et sur toute autre question importante pour la santé financière de l'Union. Les commentaires et avis du Comité des finances doivent être mis à disposition des Membres du Bureau et du Comité exécutif conformément aux Règles de travail.

30. Un Bureau administratif, sous la direction du Secrétaire général, assure la correspondance, administre les fonds, et veille à la préservation des archives de l'Union.

31. L'Union a un droit d'auteur sur tous les éléments imprimés dans ses publications, sauf disposition contraire.

VIII. CLAUSES FINALES

32. Le présent Règlement intérieur entre en vigueur le 21 Août 2012.

33. Le présent Règlement intérieur est publié en français et en anglais. À toutes fins juridiques, la version française fait autorité.

5. IAU Working Rules

I. INTRODUCTION AND RATIONALE

The Statutes of the International Astronomical Union (IAU) define the goals and organizational structure of the Union, while the Bye-Laws specify the main tasks of the various bodies of the Union in implementing the provisions of the Statutes. The Working Rules are designed to assist the membership and governing bodies of the Union in carrying out these tasks in an appropriate and effective manner. Each of the sections below is preceded by an introduction outlining the goals to be accomplished by the procedures specified in the succeeding paragraphs. The Executive Committee updates the Working Rules as necessary to reflect current procedures and to optimize the services of the IAU to its membership.

II. NON-DISCRIMINATION

The International Astronomical Union (IAU) follows the regulations of the International Council for Science (ICSU) and concurs with the actions undertaken by their Standing Committee on Freedom in the Conduct of Science on non-discrimination and universality of science (cf. §22 below)

III. NATIONAL MEMBERSHIP

The aim of the rules for applications for National Membership is to ensure that the proposed National Member adequately represents an astronomical community not already represented by another Member, and that such membership will be of maximum benefit for the community concerned (cf. Statutes §IV).

1. Applications for National Membership should therefore clearly describe the following essential conditions:

1.a. the precise definition of the astronomical community to be represented by the proposed Member;

1.b. the present state and expected development of that astronomical community;

1.c. the manner in which the proposed National Member represents this community;

1.d. whether the application is for membership on a permanent, interim, or prospective basis; and

1.e. the category in which the prospective National Member wishes to be classified (cf. Bye-Laws §25).

1.f. the process by which the National Membership annual dues will be paid promptly and in full.

2. Applications for National Membership shall be submitted to the General Secretary, who will forward them to the Executive Committee for review as provided in the Statutes.

IV. INDIVIDUAL MEMBERSHIP

Professional scientists whose research is directly relevant to some branch of astronomy are eligible for election as Individual Members of the Union (cf. Statutes §V). Individual Members are normally admitted by the Executive Committee on the proposal of a National Member. However, Presidents of Divisions may also propose individuals for membership in cases when the normal procedure is not applicable or practicable (cf. Bye-Laws §4). The present rules are intended to ensure that all applications for membership are processed on a uniform basis, and that all members are fully integrated in and contributing to the activities of the Union.

3. The term "Professional Scientist" shall normally designate a person with a doctoral degree (Ph.D.) or equivalent experience in astronomy or a related science, and whose professional activities have a substantial component of work related to astronomy.

4. National Members and Division Presidents may propose Individual Members who fall outside the category of professional scientist but who have made major contributions to the science of astronomy, e.g., through education or research related to astronomy. Such proposals should be accompanied by a detailed motivation for what should be seen as exceptions to the rule.

5. Eight months before an ordinary General Assembly, National Members and Presidents of Divisions will be invited to propose new Individual Members; these proposals should reach the General Secretary no later than five months before the General Assembly. Late proposals will normally not be taken into consideration. Proposals from Presidents of Divisions will be communicated by the IAU 3 months before the General Assembly to the relevant National Members, if any, who may add the person(s) in question to their own list of proposals.

6. National Members shall promptly inform in writing the General Secretary of the death of any Individual Member represented by them. National Members are also urged to propose the deletion of Individual Members who are no longer active in astronomy by including a written agreement of the member concerned. Such proposals should be submitted to the General Secretary at the same time as proposals for new Individual Members.

7. Proposals for membership shall include the full name, date of birth, and nationality of the candidate, postal and electronic addresses, the University, year, and subject of the M.Sc./Ph.D. or equivalent degree, current affiliation and occupation, the proposing National Member or Division, the Division(s) and/or Commission(s) which the candidate wishes to join, and any further detail that might be relevant.

8. The standing Membership Committee advises the Executive Committee on its behalf on matters related to the admission of Individual Members (Statute 13f.).

8.a Three months before a General Assembly, the Membership Committee shall prepare a list of at least 10 Individual Members of the Union who accept to serve on the Committee for the next triennium if elected, including a nominee for Chair. The General Secretary shall forward this list to the National Members and invite additional nominations from them with a deadline of one month before the General Assembly. The Membership Committee shall verify that the resulting slate complies with the rules in Statutes 13 and with general principles of scientific, geographical and gender balance. Members shall not normally serve more than two consecutive terms, and it is desirable that roughly half of the members are replaced at each election. The Chair of the Membership Committee shall present the resulting slate of nominations to the National Members together with the report of the Committee on the previous triennium at the beginning of the General Assembly, for final election at its closing session.

9. The General Secretary shall submit all proposals for Individual Membership to the Membership Committee for review, consolidated into two lists:

9.a one containing all proposals by National Members; and

9.b one containing all proposals by Presidents of Divisions, in accordance with Bye-Law §4.

10. The Membership Committee shall examine all proposals for individual membership and advise the Executive Committee on the proposals for individual membership.

11. In exceptional cases, the Executive Committee may, on the proposal of a Division, admit an Individual Member between General Assemblies. Such proposals shall be prepared as described above (cf. §2) and submitted with a justification of the request to bypass the normal procedure. The Executive Committee shall consult the Nominating Committee or relevant National Member before approving such exceptions to the normal procedure.

12. The General Secretary shall maintain updated lists of all National and Individual Members, and shall make these available to the membership in electronic form. The procedures for dissemination of these lists shall be set by the General Secretary in such a way that the membership directory be properly protected against unintended or inappropriate use.

V. RESOLUTIONS OF THE UNION

Traditionally, the decisions and recommendations of the Union on scientific and organizational matters of general and significant importance are expressed in the Resolutions of the Union. In order for such Resolutions to carry appropriate weight in the international community, they should address astronomical matters of significant impact on the international society, or matters of international policy of significant importance for the international astronomical community as a whole.

Resolutions should be adopted by the Union only after thorough preparation by the relevant bodies of the Union. The proposed resolution text should be essentially complete before the beginning of the General Assembly, to allow Individual and National Members time to study them before discussion and debate by the General Assembly. The following procedures have been designed to accomplish this:

13. Proposals for Resolutions to be adopted by the Union may be submitted by a National Member, by the Executive Committee, a Division, a Commission or a Working Group. They should address specific issues of the nature described above, define the objectives to be achieved, and describe the action(s) to be taken by the Officers, Executive Committee, or Divisions to achieve these objectives.

14. Resolutions proposed for vote by the Union fall in three categories as set out in Article 14 of the Union's Statutes:

14.a. Resolutions with implications for the budget of the Union (Statute 14a); or

14.b. Resolutions affecting the administration of the Union but without financial implications (Statute 14b).

14.c. Resolutions of a primarily scientific nature (Statute 14).

Proposals for Resolutions should be submitted on standard forms appropriate for each type, which are available from the IAU Secretariat. They may be submitted in either English or French and will be discussed and voted upon in the original language. Upon submission each proposed Resolution is posted on the Union web site. When the approved Resolutions are published, a translation to the other language will be provided.

15. Resolutions with implications for the budget of the Union must be submitted to the General Secretary at least nine months before the General Assembly in order to be taken into account in the budget for the impending triennium. All other Resolutions must be submitted to the General Secretary six months (Bye-Laws 7c and 7d) before the beginning of the General Assembly. The Executive Committee may decide to accept late proposals in exceptional circumstances.

16. Before being submitted to the vote of the General Assembly, proposed Resolutions will be examined by the Executive Committee, Division Presidents, and by a Resolutions Committee, which is nominated by the Executive Committee. The Resolutions Committee consists of at least three members of the Union, one of whom should be a member of the Executive Committee, and one of whom should be a continuing member from the previous triennium. It is appointed by the General Assembly during its final session and remains in office until the end of the following General Assembly.

17. The Resolutions Committee will examine the content, wording, and implications of all proposed Resolutions promptly after their submission. In particular, it will address the following points:

i. suitability of the subject for an IAU Resolution;

ii. correct and unambiguous wording;

iii. consistency with previous IAU Resolutions.

The Resolutions Committee may refer a Resolution back to the proposers for revision or withdrawal if it perceives significant problems with the text, but can neither withdraw nor modify its substance on its own initiative. The Resolutions Committee advises the Executive Committee whether the subject of a proposed Resolution is primarily a matter of policy or primarily scientific. The Resolutions Committee will also notify the Executive Committee of any perceived problems with the substance of a proposed Resolution.

18. The Executive Committee will examine the substance and implications of all proposed Resolutions. Proposed Resolutions shall be published in the General Assembly Newspaper before the final session, and shall state if the Resolution is open to electronic voting after the General Assembly. The Resolutions Committee will present the proposals during a plenary session of the General Assembly with its own recommendations, and those of the Executive Committee, if any, for their approval or rejection. A representative of the body proposing the Resolution will be given the opportunity to defend the Resolution in front of the General Assembly, after which a general discussion shall take place

19. Resolutions with implications for the budget of the Union are voted upon by the National Members during the final plenary session of the General Assembly. Other resolutions may be voted upon by the National Members or by Individual Members as appropriate according to the Statutes of the Union by correspondence after the General Assembly. The Union will facilitate electronic discussion of all Resolutions on the Union website in advance of a vote either at the General Assembly or electronically (Statutes 15a. and 15.b.).

VI. EXTERNAL RELATIONS

Contacts with other international scientific organizations, national and international public bodies, the media, and the public are increasing in extent and importance. In order to maintain coherent overall policies in matters of international significance, clear delegation of authority is required. Part of this is accomplished by having the Union's representatives in other scientific organization appointed by the Executive Committee (cf. Bye-Laws §15). Supplementary rules are given in the following.

20. Representatives of the Union in other scientific organizations are appointed by the Executive Committee upon consultation with the Division(s) in the field(s) concerned.

21. In other international organizations, e.g. in the United Nations Organization, the Union is normally represented by the General Secretary or Assistant General Secretary, as decided by the Executive Committee.

22. The Union strongly supports the policies of the International Council for Science (ICSU) as regards the freedom and universality of science. Participants in IAU sponsored activities who feel that they may have been subjected to discrimination are urged, first, to seek clarification of the origin of the incident, which may have been due to misunderstandings or to the cultural differences encountered in an international environment. Should these attempts not prove successful, contact should be made with the General Secretary who will take steps to resolve the issue.

23. Public statements that are attributed to the Union as a whole can be made only by the President, the General Secretary, or the Executive Committee. The General Secretary may, in consultation with the relevant Division, appoint Individual Members of the Union with special expertise in questions that attract the attention of media and the general public as IAU spokespersons on specific matters.

VII. FINANCIAL MATTERS

The great majority of the Union's financial resources are provided by the National Members, as laid out in the Statutes §XII and Bye-Laws §VII. The purpose of the procedures described below is twofold: (i) to provide the best possible advice and guidance to the General Secretary and Executive Committee in planning and managing the Union's financial affairs, and (ii) to provide National Members with a mechanism for continuing input to and oversight over these affairs between and in preparation for the General Assemblies. The procedures adopted to accomplish this are as follows:

24. At the end of each of its final sessions the General Assembly appoints a Finance Sub-Committee of 5-6 members, including a Chair.

24.a. Three months before a General Assembly, the Finance Committee shall prepare a list of at least 10 Individual Members of the Union who accept to serve on the Committee for the next triennium if elected, including a nominee for Chair. The General Secretary shall forward this list to the National Members and invite additional nominations from them with a deadline of one month before the General Assembly. The Finance Committee shall verify that the resulting slate complies with the rules in Statutes 13 and with general principles of scientific, geographical and gender balance.

Members shall not normally serve more than two consecutive terms, and it is desirable that roughly half of the members are replaced at each election. The Chair of the Finance Committee shall present the resulting slate of nominations to the National Members together with the report of the Committee on the previous triennium at the beginning of the General Assembly, for final election at its closing session. The Finance Committee remains in office until the end of the next General Assembly (cf. Statutes §13.d.) and cooperates with the National Members, Executive Committee and General Secretary in the following manner:

24.b. After the end of each year the General Secretary will call for a legal audit of the accounts by a properly licensed, external auditor. The auditor will make a report addressed to the General Assembly. The General Secretary provides the Finance Sub-Committee with the auditor's report and summary reports covering the financial performance of the Union as compared to the approved budget, together with an analysis of any significant departures, and information on any Executive Committee approvals of budget changes. Upon receipt of the above reports from the General Secretary, the Finance Sub-Committee examines the accounts of the Union in the light of the corresponding budget and any relevant later decisions by the Executive Committee. It reports its findings and recommendations to the Executive Committee at its next meeting. The Finance Sub-Committee may at any time, at the request of the Executive Committee or the General Secretary, or on its own initiative, advise the General Secretary and/or the Executive Committee on any aspect of the Union's financial affairs.

24.c. Towards the end of the year preceding that of a General Assembly, the General Secretary shall submit a preliminary draft of the budget for the next triennium to the Finance Committee for review. The draft budget, updated as appropriate following the comments and advice of the Finance Committee, shall be submitted to the Executive Committee for approval together with the report of the Finance Committee and shall be sent to the National Members as a draft budget as stipulated in Bye-Law 7b. The final budget proposal as approved by the Executive Committee shall be submitted to the National Members with a statement of the views of the Finance Committee on the proposal.

24.d. Before the first session of a General Assembly, the Finance Sub-Committee shall submit a report, including the auditor's reports, to the Executive Committee on its findings and recommendations concerning the development of the Union's finances over the preceding triennium. The Finance Sub-Committee shall also prepare, in consultation with the National Members, a slate of candidates for the composition of the Finance Sub-Committee in the next triennium, preferably providing a balance between new and continuing members.

24.e. The report of the Finance Committee, together with the audited detailed accounts and the earlier comments on the proposed budget for the next triennium, will form a suitable basis for the discussions of the Finance Committee leading to its recommendations to the General Assembly concerning the approval of the accounts for the previous triennium and the budget for the next triennium, as well as the new Finance Sub-Committee to serve during that period.

25. The General Secretary is responsible for managing the Union's financial affairs according to the approved budget (cf. Bye-Laws §28).

25.a. In response to changing circumstances, the Executive Committee may approve such specific changes to the annual budgets as are consistent with the intentions of the General Assembly when the budget was approved.

25.b. Unless authorized by the Executive Committee, the General Secretary shall not approve expenses exceeding the approved budget by more than 10% of any corresponding major budget line or 2% of the total budget in a given year, whichever is larger. This restriction does not apply in cases when external funding has been provided for a specific purpose, e.g. travel grants to a General Assembly.

25.c. Unless specifically identified in the approved budget, contractual commitments in excess of €50,000, or with performance terms in excess of 3 years require the additional approval of the Union President.

26. The National Representatives, in approving the accounts for the preceding triennium, discharge the General Secretary and the Executive Committee of liability for the period in question.

VIII. RULES OF PROCEDURE FOR THE EXECUTIVE COMMITTEE

The Executive Committee must respond quickly to events and thus it needs to be able to have discussions and take decisions on a relatively short timescale and without meeting in person. The following rules, as required by Bye-Law 14, are designed to facilitate EC action in a flexible manner, while giving such decisions the same legal status as those taken at actual physical meetings.

27. The Executive Committee should meet in person at least once per year. In years of a General Assembly it should meet in conjunction with and at the venue of the General Assembly. In other years, the Executive Committee decides on the date and venue of its regular meeting. The meetings of the Executive Committee are chaired by the President or, if the President is unavailable, by the President Elect or by one of the Vice-Presidents chosen by the Executive Committee to serve in this capacity.

28. The date and venue of the next regular meeting of the Executive Committee shall be communicated at least six months in advance to all its members and the Advisors, and to all Presidents of Divisions. Any of these persons may then propose items for inclusion in the Draft Agenda of the meeting before the date posted on the IAU Deadlines page.

29. Outgoing and incoming Presidents of Divisions are invited to attend all non-confidential sessions of the outgoing and incoming Executive Committee, respectively, in the year of a General Assembly. The President will invite Presidents of Divisions to attend the meetings of the Executive Committee in the years preceding a General Assembly. Division Presidents attend these sessions with speaking right, but do not participate in any voting.

30. The Executive Committee may take official decisions if at least half of its members participate in the discussion and vote on an issue. Decisions are taken by a simple majority of the votes cast. In case of an equal division of votes, the Chair's vote decides the issue. Members who are unable to attend may, by written or electronic correspondence with the President before the meeting, authorize another member to vote on her/his behalf or submit valid votes on specific issues.

31. If events arise that require action from the Executive Committee between its regular meetings, the Committee may meet by teleconference or by such electronic or other means of correspondence as it may decide. In such cases, the Officers shall submit a clear description of the issue at hand, with a deadline for reactions. If the Officers propose a specific decision on the issue, the decision shall be considered as approved unless a majority of members vote against it by the specified deadline. In case of a delay in communication, or if the available information is considered insufficient for a decision, the deadline shall be extended or the decision deferred until a later meeting at the request of at least two members of the Executive Committee.

32. The Officers of the IAU should, as a rule, meet once a year at the IAU Secretariat in order to discuss all matters of importance to the Union. The other members of the Executive Committee and the Division Presidents shall be invited to submit items for discussion at the Officers' Meetings and shall receive brief minutes of these Meetings.

33. Should any member of the Executive Committee have a conflict of interest on a matter before the Executive Committee that might compromise their ability to act in the best interests of the Union, they shall declare their conflict of interest to the Executive Committee, and such conflict shall be recorded by the Secretariat. The remaining members of the Executive Committee determine the appropriate level of participation in such issues for members with a potential conflict of interest.

IX. SCIENTIFIC MEETINGS AND PUBLICATIONS

Meetings and their proceedings remain a major part of the activities of the Union. The purpose of scientific meetings is to provide a forum for the development and dissemination of new ideas, and the proceedings are a written record of what transpired.

34. The General Secretary shall publish in the Transactions and on the IAU web site rules for scientific meetings organized or sponsored by the Union.

35. The proceedings of the General Assemblies and other scientific meetings organized or sponsored by the Union shall, as a rule, be published. To ensure prompt publication of Proceedings of IAU Symposia and Colloquia, the Assistant General Secretary is authorized to oversee the production of the material for the Proceedings. The Union shall publish an Information Bulletin at regular intervals to keep Members informed of current and future events in the Union. The Union shall also publish a more informal, periodic Newsletter which it distributes electronically to its members. The Executive Committee decides on the scope, format, and production policies for such publications, with due regard to the need for prompt publication of new scientific results and to the financial implications for the Union. At the present time, publications are in printed and in electronic form.

36. Divisions, Commissions, and Working Groups shall, with the approval of the Executive Committee, be encouraged to issue Newsletters or similar publications addressing issues within the scope of their activity.

X. TERMS OF REFERENCE FOR DIVISIONS

The Divisions are the scientific backbone of the IAU. They have a main responsibility for monitoring the scientific and international development of astronomy within their subject areas, and for ensuring that the IAU will address the most significant issues of the time with maximum foresight, enterprising spirit, and scientific judgment. To fulfill this role IAU Divisions should maintain a balance between innovation and continuity. The following standard Terms of Reference have been drafted to facilitate that process, within the rules laid down in the Statutes §X and the Bye-Laws §V.

37. As specified in Bye-Law 18, the scientific affairs of the Division are conducted by an Organizing Committee of up to 12 members of the Division, headed by the Division President, Vice-President, and Secretary. Thus, all significant decisions of the Division require the approval of the Organizing Committee, and the President and Vice-President are responsible for organizing the work of the Committee so that its members are consulted in a timely manner. Contact information for the members of the Organizing Committee shall be maintained at the Division web site.

Unless agreed otherwise by the Executive Committee on a case by case basis, the President of a Division cannot be President of another Division or of a Commission, or be Chair of a Working Group.

38. Individual Members of the Union are admitted to membership in a Division by its Organizing Committee (cf. Bye-Laws §18). Individual Members active within the field of activity of the Division and interested in contributing to its development should contact the Division Secretary, who will consult the Organizing Committee on the admission of the candidates.

38.a. The Division Secretary shall maintain a list of Division members for ready consultation by the community, including their Commission memberships if any. Updates to the list shall be provided to the IAU Secretariat on a running basis.

38.b. Members may resign from a Division by so informing the Division Secretary.

38.c. In the event of a Division being newly formed, Individual Members can themselves elect to join the Division. Before the General Assembly following that at which the new Division was created its Organising Committee shall scrutinise and confirm the Division membership.

39. The effectiveness of the Division relies strongly on the scientific stature and dedication of its President and Vice-President to the mission of the Division. The Executive Committee, in proposing new Division Presidents and Vice-Presidents for election by the General Assembly, will rely heavily on the recommendations of the Organizing Committee of the Division. In order to prepare a strong slate of candidates for these positions, and for the succession on the Organizing Committee itself, the following procedures shall normally apply:

39.a. Candidates are proposed and selected from the membership of the Division on the basis of their qualifications, experience, and stature in the fields covered by the Division. In addition, the Organizing Committees should have proper gender balance and broad geographical representation.

39.b. At least six months before a General Assembly, the Organizing Committee submits to the membership of the Division a list of candidates for President, Vice-President (for which there should be at least two persons willing to serve), Secretary, and the Organizing Committee for the next triennium. The Organizing Committee requests nominations from the entire membership in preparing this list, and then conducts a vote, normally electronically, among all Division members for the above offices, the results of which are reported to the General Secretary at least three months before the General Assembly. The Vice-President is normally nominated to succeed the President. The outgoing President participates in the deliberations of the new Organizing Committee in an advisory capacity.

39.c. If more names are proposed than there are positions to be filled on the new Organizing Committee, the outgoing Organizing Committee devises the procedure by which the requisite number of candidates is elected by the membership. The resulting list is communicated to the General Secretary at least two months before the General Assembly. The General Secretary may allow any outstanding issues to be resolved at the business meeting of the Division during the General Assembly. If for any reason the Organizing Committee has not been able to arrange for the election of new officers and an Organizing Committee by two months before the GA, the EC will nominate a VP and Organizing Committee at its first General Assembly meeting.

39.d. A member of the Organizing Committee normally serves a maximum of two terms, unless elected Vice-President of the Division in her/his second term. Presidents may serve for only one term.

39.e. The Organizing Committee decides on the procedures for designating the Division Secretary, who maintains the web site, records of the business and membership of the Division, and other rules for conducting its business by physical meetings or by correspondence.

39.f. In the event of a newly formed Division, paragraphs 39a - 39c do not apply. The Executive Committee shall consult the Organizing Committees of the relevant predecessor Divisions on possibl candidates for President and Vice-President of the new Division for the next triennium. The Executive Committee shall select the names to be proposed to the Generals Assembly for election.

39.g. As soon as possible after their election at a General Assembly, the President and Vice-President of the new Division shall request nominations to the Organising Committee from the membership of the Division and then conduct a vote among Division members, the results of which are reported to the General Secretary. The Organising Committee elects a Secretary from its membership.

40. A key responsibility of the Organizing Committee is to maintain an internal organization of Commissions and Working Groups in the Division which is conducive to the fulfillment of its mission. The Organizing Committee shall take the following steps to accomplish this task in a timely and effective manner:

40.a. Within the first year after a General Assembly - with the business meeting of the Commission at the General Assembly itself as a natural starting point - the Organizing Committee shall discuss with its Commissions, and within the Organizing Committee itself, if changes in its Commission and Working Group structure may enable it to accomplish its mission better in the future. As a rule, Working Groups should be created (following the rules in Bye-Law 21 and Bye-Law 23) for new activities that are either of a known, finite duration or are exploratory in nature. If experience, possibly from an existing Working Group, indicates that a major section of the Division's activities require a coordinating body for a longer period (a decade or more), the creation of a new Commission may be in order.

40.b. Whenever the Organizing Committee is satisfied that the creation of a new Working Group or Commission is well motivated, it may take immediate action as specified in Bye-Law 21 or Bye-Law 23. In any case, the Organizing Committee submits its complete proposal for the continuation, discontinuation, or merger of its Commissions and Working Groups to the General Secretary at least three months before the next General Assembly.

40.c. The President and Organizing Committee maintain frequent contacts with the other IAU Divisions to ensure that any newly emerging or interdisciplinary matters are addressed appropriately and effectively.

XI. TERMS OF REFERENCE FOR COMMISSIONS

The role of the Commissions is to organize the work of the Union in specialized subsets of the fields of their parent Division(s), when the corresponding activity is judged to be of considerable significance over times of a decade or more. Thus, new Commissions may be created by the Executive Committee with the agreement of all the Divisions when fields emerge that are clearly in sustained long-term development and where the Union may play a significant role in promoting this development at the international level. Similarly, Commissions may be discontinued by the Executive Committee upon the recommendation of the parent Division when their work can be accomplished effectively by the parent Division. In keeping with the many-sided activities of the Union, Commissions may have purely scientific as well as more organizational and/or interdisciplinary fields. They will normally belong and report to one of the IAU Divisions, but may be common to two or more Divisions. The following rules apply if a Division has more than one Commission.

41. The activities of a Commission are directed by an Organizing Committee of 4-8 members of the Commission, headed by a Commission President and Vice-President (cf. Bye-Laws §22). A member of the Organizing Committee normally serves a maximum of two terms, unless elected Vice-President of the Commission in her/his second term. Presidents may serve for only one term. All members of the Organizing Committee are expected to be active in this task, and are to be consulted on all significant actions of the Commission. The Organizing Committee appoints a Commission Secretary who maintains the records of the membership and activities of the Commission in co-operation with the Division Secretary and the IAU Secretariat. Contact information for the members of the Organizing Committee shall be maintained at the Commission web site.

Unless agreed otherwise by the Executive Committee on a case by case basis, the President of a Commission cannot be President of a Division or of another Commission, or be Chair of a Working Group.

42. Individual Members of the Union, who are active in the field of the Commission and wish to contribute to its progress, are admitted as members of the Commission by the Organizing Committee. Interested Members should contact the Commission Secretary, who will bring the request before the Organizing Committee for decision. Members may resign from the Commission by notifying the Commission Secretary. Before each General Assembly, the Organizing Committee may also decide to terminate the Commission membership of persons who have not been active in the work of the Commission; the individuals concerned shall be informed of such planned action before it is put into effect. The Commission Secretary will report all changes in the Commission membership to the Division Secretary and the IAU Secretariat.

43. At least six months before a General Assembly, the Organizing Committee submits to the membership of the Commission a list of candidates for President, Vice-President (for which there should be the names of two persons willing to serve), the Organizing Committee, and heads of Program Groups for the next triennium. The Organizing Committee requests nominations from the entire membership in preparing this list, and then conducts a vote, normally electronically, among all the members for the above offices, the results of which are reported to the General Secretary at least three months before the General Assembly. The Vice-President is normally nominated to succeed the President. The outgoing President participates in the deliberations of the new Organizing Committee in an advisory capacity. If more names are proposed than available elective positions, the outgoing Organizing Committee devises the procedure by which the requisite number of candidates is elected by the membership. The resulting list is submitted to the Organizing Committee of the parent Division(s) for approval before the end of the General Assembly. Members of the Organizing Committee normally serve a maximum of two terms, unless elected Vice-President of the Commission. Presidents may serve for only one term.

44. At least six months before each General Assembly, the Organizing Committee shall submit to the parent Division(s) a report on its activities during the past triennium, with its recommendation as to whether the Commission should be continued for another three years, or merged with one or more other Commissions, or discontinued. If a continuation is proposed, a plan for the activities of the next triennium should be presented, including those of any Working Groups which the Commission proposes to maintain during that period.

45. The Organizing Committee decides its own rules for the conduct of its business by physical meetings or (electronic) correspondence. Such rules require approval by the Organizing Committee of the parent Division(s).

46. The procedural rules applying to the establishment of a new Division shall also apply to the establishment of a new Commission. Where there is no "relevant predecessor Commission(s)" the parent Division(s) shall submit to the potential membership of the new Commission a list of candidates for President, Vice-President and Organising Committee for the next triennium.

XII. SPECIAL NOMINATING COMMITTEE

47. Approximately six months before the start of the General Assembly the General Secretary shall invite Division Presidents and the members of the Executive Committee to nominate potential members of the SNC with a deadline of 3 months before the General Assembly. The Executive Committee shall prepare a list of candidates in consultation with the Membership Committee for appointment at the final Business Session of the General Assembly. The SNC, once appointed, shall elect its own Chair.

Transactions IAU, Volume XXX
Proc. XXX IAU General Assembly, August 2018
Debra Elmegreen, ed.
doi:10.1017/S1743921323005677

CHAPTER IX

THIRTIETH GENERAL ASSEMBLY

NEW MEMBERS AND DECEASED MEMBERS PRESENTED AT THE GENERAL ASSEMBLY

1. Report of the IAU Membership Committee on the last Triennium (2015–2018)

Members of the IAU Membership Committee:

Sona EHLEROVA (Czech Republic)
William HARRIS (Canada)
Myungshin IM (Rep. Korea)
Lex KAPER (The Netherlands)
Daniela LAZZARO (Brazil)
Helmut O. RUCKER (Austria), Chair of the Committee
Ramotholo SEFAKO (South Africa)
David SODERBLOM (USA)

For the reviewing process all applications have been shared among the Committee members by nations which are - at least to some extent - close and related to the nationality of the respective Committee members.

Statistics

The overall IAU Membership Applications submitted to IAU, cleared up (i.e., removed) duplicates, comprised in number 1314 applications (Line 2 in Table 1). Out of this number the total number of IAU Individual Membership applications is 919 applications (Line 3 in Table 1) and the total number of IAU Junior Membership applications is 395 applications (Line 4 in Table 1).

A detailed statistics with regard to gender was not possible since this information was not visually provided in the online application. Several tens of applications have been transferred from IAU Individual to IAU Junior Membership due to the constraints posed on the Individual Membership, vice versa (i.e. transfer from Junior to Individual Membership) regarded only a few applications. This transfer process (as well as the rejections) addresses the conditions under which an IAU Individual or Junior Membership can be obtained. The numbers given in Table 1 clearly demonstrate that the introduction of the IAU Junior Membership was a success.

Table 1. Membership Applications

1	Total applications (Individual + Junior) including duplicates:	1356
2	Total applications (Individual + Junior) without duplicates:	1314
3	Total Individual Membership applications:	919
4	Total Junior Membership applications:	395
5	Number of Individual Membership applications accepted:	888
6	Number of Individual Membership applications rejected:	31
7	Number of Junior Membership applications accepted:	352
8	Number of Junior Membership applications rejected:	43

Online Applications

The application for both Individual and Junior Membership has been performed via online process and this procedure in general worked out quite well. However, some important improvements should be implemented in order to obtain a better performance both in the procedure of online application as well as in the reviewing process. In the following some recommendations are listed, in categories of major (A) and minor importance (B).

A1) Clear rules of conditions for the respective memberships which should be emphasized to the attention of the respective NCAs.

A2) Clear rules on the responsibility of application, i.e., what is relevant for the affiliation: the nation?, the citizenship?, the working institution and present position?

A3) Clear defined deadlines for the online application, for the NCA comments, for the Membership Reviewing process.

A4) The online form should define obligatory fields. When these fields are not filled with information then the application should not get finalized. In most cases of incomplete applications the "List of publications" and the "Present position" was missing.

B1) The online form poses the question: "Have you authored or been the principal co-author of refereed papers?" and "Application for IAU Membership:" All answers were "no" which does not make sense.

B2) Letters of Reference should be written in English (some have been written in Chinese language).

B3) Female applications may contain the maiden name if the name has been changed (relevant for publication check).

Acknowledgements

It is to the pleasure of the signatory to fully acknowledge the excellent work of all members of the IAU Membership Application Committee. The signatory also wants to emphasise the great support provided by both Madeleine Smith-Spanier and Piero Benvenuti having done speedy replies on requests and corresponding transfer actions.

The List of New Members

Below the names of the new members are given, recommended by the Membership Committee for admission at the General Assembly. Those nominated by the Division Presidents are presented first, followed by the majority of successful applicants, which were nominated by National Members.

NEW MEMBERS ADMITTED BY THE XXXth GENERAL ASSEMBLY

VIENNA, 20–31 AUG. 2018

The following lists give the names of the 352 new Junior Members and of the 888 Individual Members admitted at the XXXth General Assembly, ordered by National Member.

For a complete list of IAU members, please consult the IAU Directory:
`http://www.iau.org/administration/membership/individual/`.

For a complete list of IAU National Members (member countries), please visit:
`http://www.iau.org/administration/membership/national/`.

1. New Individual Members nominated by Division Presidents

Division D Tudose, Valeriu (Romania)

Division E Derouich, Moncef, (Tunisia)

Division H Asa'd, Randa, (United Arab Emirates)

2. New Junior Members listed by National Member

Australia

Murphy, Simon
Casey, Andrew
Janowiecki, Steven
Fitzgerald, Michael
Sansom, Eleanor
Wong, Graeme

Austria

Boro Saikia, Sudeshna

Belgium

MERLE, Thibault
Mossoux, Enmanuelle

Brazil

Barbosa, Carlos
Galli, Phillip

Bulgaria

Iliev, Lubomir

Canada

Rajabi, Fereshteh
Drout, Maria
Henault-Brunet, Vincent
Naud, Marie-Eve
Rousseau-Nepton, Laurie
Brown, Toby
Cridland, Alexander
van der Marel, Nienke
Ngo, Henry

Chile

Espinoza, Néstor

China, Nanjing

LIU, Zhiyong
ZHOU, Xia
SHEN, Jinhua
AN, Ning
LI, Ying
REN, Yuan
WEN, Haikun
LI, Zhenwei
CHEN, Xiaoyan
YANG, Shu-Hua
TANG, Jianfei
KANG, Zhe
XU, Yan
YUAN, Qiang
ZHANG, Yu
NIE, Jundan
JIN, Zhiping
LI, Weichao
ZHANG, Kai
ZHANG, Ming-Jiang
WANG, Wei
YANG, Wenbo
ZHANG, Hailong
HOU, Junfeng
YAN, Xiaoli
WU, Wenjun
GUO, Jianhua
ZHANG, Xiao-Jun
FENG, Li
XU, Yidong
WENG, Shan-Shan
BAI, Hua
WEN, guanyu
WEI, Jun-Jie
TONG, Minglei
SHI, Jianchun
SU, Yang
SONG, Yihan
HOU, Wen
HUANG, Jing
DENG, Xue-Mei
LIU, Ya
LIU, Jun
FAN, Dongwei
LIU, Jinzhong
ZHANG, Jujia
SU, Yingna
LI, Yinbi
CUI, Lang
GE, Hongwei
LI, Xiaoyan

Colombia

Mojica-Gómez, Sindy

Czech Republic

Klement, Robert
Vojáček, Vlastimil
Maryeva, Olga

Denmark

Jaisawal, Gaurava Kumar

Egypt

Elshaer, Ahmed
Mohamed, Walid
El-Nazer, Mostafa
Ahmed, Ahmed
zead, ibrahim
Abulwfa, Ahmed
Abdelbar, Ahmed
Ibrahim, Ahmed Hafez
Hussein, Hussein
Elkhamisy, Mohammad
Abdelkawy, Ali

Ethiopia

Mekuria, Remudin

Finland

Kuncarayakti, Hanindyo
Nättilä, Joonas
Kangas, Tuomas
Watkins, Aaron
Laine, Jarkko

France

AL-EDHARI, Ali
Fragkoudi, Francesca
Giammichele, Noemi
Sorce, Jenny
Martinet, Nicolas
Suer, Terry-Ann
Danielski, Camilla
Rousseau, Batiste
Gavilan, Lisseth
Kebe, Fatoumata

Germany

Dzib Quijano, Sergio
Myserlis, Ioannis
Oldengott, Isabel
Chatzistergos, Theodosios
Kee, Nathaniel
TIburzi, Caterina
Anders, Friedrich
Facchini, Stefano
Colombo, Dario
McKee, James
Cantalloube, Faustine
Zajacek, Michal
Deng, Xinping
Basu, Aritra
Antoniadis, John
Hoffmann, Susanne
Lee, Min-Young
Ortiz Leon, Gisela
Sander, Andreas

Hungary

Varga-Verebelyi, Erika
Kun, Emma
Cruz-Saenz de Miera, Fernando
Kriskovics, Levente
Kadam, Kundan
Nagy, Andrea
Dobos, Vera
Belucz, Bernadett

India

Pawar, Devraj
Sharma, Mohit
Sachdeva, Sonali
Pandge, Mahadev
Padmanabhan, Hamsa
Mohanty, Pravata
Kumar, Brajesh
Kollamparambil Paul,
Arun, Babu
Chaturvedi, Priyanka

Indonesia

Priyatikanto, Rhorom

Israel

Faigler, Simchon
Kagan, Daniel
Freundlich, Jonathan
Tal-Or, Lev
Tiwari, Prabhakar
Globus, Noemie
Malamud, Uri

Italy

Sánchez Ramírez, Rubén
Rossi, Andrea
Giannetti, Andrea
Zambon, Francesca
Bruni, Gabriele
Murabito, Mariarita
Varano, Stefania
Paggi, Alessandro
Rivilla, Víctor
Rygl, Kazi
Piccinni, Ornella
IEVA, SIMONE
Pilia, Maura

Japan

TAKASAO, Shinsuke
Zahorecz, Sarolta

Korea, Republic of

CHO, Hyejeon

Lithuania

Klevas, Jonas
Drazdauskas, Arnas
Čepas, Vytautas

Mexico

López Valdivia, Ricardo
Ibarra-Medel, Hector
Hueyotl-Zahuantitla, Filiberto
VILLICANA-PEDRAZA, ILHUIYOLITZIN
Bartali, Roberto
Ortega-Minakata, René
Zavala, Jorge

Netherlands

Ellerbroek, Lucas
Nanayakkara, Themiya
Cai, Maxwell Xu
Todorov, Kamen
Silva, Marta
Toonen, Silvia
Kannawadi Jayaraman, Arun
Murillo Mejias, Nadia
Massari, Davide
Burns, Ross
Chhotray, Atul
Beuchert, Tobias
Contreras, Yanett
de Boer, Jozua
Hacar, Alvaro
Immer, Katharina
Intema, Huib
Lena, Davide
Marcote, Benito
Mushtukov, Alexander
Moscibrodzka, Monika
Moss, Vanessa
Oman, Kyle
Patruno, Alessandro
Posti, Lorenzo
Japelj, Jure
Rab, Christian
Ribeiro, Bruno

New Zealand

Alexander, Jordan

Norway

Froment, Clara
Szydlarski, Mikolaj Marcin
Gronke, Max
Laursen, Peter
de Jorge Henriques, Vasco
Manuel

Poland

Karska, Agata
Szewczuk, Wojciech
Czart, Krzysztof
Ratajczak, Milena
Zieliński, Paweł

Portugal

Santos, Angela
Gomes, Nuno

Russian Federation

Nosov, Evgeny
Baklanova, Dilyara
Babina, Elena
Artemenko, Svetlana
Antonyuk, Oksana
Perepelitsyna, Julia
Kartashova, Anna
Svinkin, Dmitry
Uklein, Roman
Abunin, Artem
Murga, Maria
Levkina, Polina
Tarasov, Sergei
Klimenko, Viacheslav
Malanchev, Konstantin
Puzin, Vasilii
Sitnova, Tatyana
Vinokurov, Alexander
Kochina, Olga
Gabdeev, Maksim
Zagainova, Iuliia
Khabibullin, Ildar
Goglichidze, Oleg
Razdoburdin, Dmitry
Kolbin, Alexander
Vavilov, Dmitrii
Shikhovtsev, Artem

Serbia, Republic of

Vudragović, Ana
Smole, Majda
Vučetić, Milica
Knezevic, Sladjana
Marceta, Dusan
Ćiprijanović, Aleksandra
Stojanović, Milan
Martinović, Nemanja
Martinović, Mihailo

South Africa

Kasai, Eli,
Randriamanakoto, Zara
Kotze, Enrico
Molaro, Margherita
Kuhn, Rudolf
Erasmus, Nicolas
Mogotsi, Moses
Groenewald, Danièl
Randriamampandry, Solohery

Spain

Belda Palazon, Santiago
Beck, Paul
Ribas, Salvador
Velilla-Prieto, Luis
Jones, Michael

Switzerland

Drozdovskaya, Maria
Chan, James
Sterken, Veerle
Bozzo, Enrico
Blazek, Jonathan
Raichoor, Anand
Schober, Jennifer
Marschall, Raphael

Tajikistan

Khamroev, Umedzhon,

Thailand

Awiphan, Supachai
Tanakul, Nahathai
Pongkitivanichkul, Chakrit
Chainakun, Poemwai

Turkey

Şener, H. Tuğça
Pektas, Riza

Ukraine

Klyuyeva, Antonina
Yerin, Serge
Vasylkivskyi, Yevhen
Kravtsov, Ihor
Vasylieva, Iaroslavna
Shubina, Olena
Shevchuk, Mykola
Koshmak, Ihor
Khoptar, Alina
Mylostna, Krystyna
Berdina, Liudmyla
Guliyev, Rustam
Pomazan, Anton
Dobrycheva, Daria
Tkachenko, Mariia
Ponomarenko, Vasyl
Kulichenko, Mykola
Kaliuzhnyi, Mykola
Torbaniuk, Olena

United Kingdom

Vincenzo, Fiorenzo
Wienen, Marion
Onions, Julian
Tazzari, Marco
Davies, Claire
Morabito, Leah
Holdsworth, Daniel
Battino, Umberto
Sànchez de Miguel, Alejandro
Rolfe, Samantha
Kuchner, Ulrike

United States

Egeland, Ricky
Pauly, Tyler
Ye, Quan-Zhi
Pacucci, Fabio
Gonzales, Eileen
Coughlin, Eric
Hamers, Adrian
Ali, Aamir
Kuhn, Michael
Lam, Michael
Li, Ting
MacLeod, Morgan
Libralato, Mattia
Godfrey, Paige
Pradhan, Pragati
Brewer, John
Pope, Benjamin
Lewandowska, Natalia
Hirschauer, Alec
Xu, Zhilei
Rivera Sandoval, Liliana
Rau, Gioia
David-Uraz, Alexandre
Rajan, Abhijith
Li, Miao
Goldman, Steven

3. New Individual Members listed by National Member

Argentina

Bersten, Melina
Cabello, Iballa
Cañada Assandri, Marcela
Correa-Otto, Jorge
Scóccola, Claudia
Díaz, Rodrigo
Giuppone, Cristian
Gramajo, Luciana
Guilera, Octavio
Lares, Marcelo
Palma, Tali
Camino, Nestor

Australia

Slagmolen, Bram
Kafle, Prajwal
Lagos, Claudia
Groves, Brent
Wong, Oiwei
Müller, Bernhard
Price, Danny
Lazendic-Galloway, Jasmina
Ryan-Weber, Emma
Veitch, Peter
Sinha, Manodeep
Waite, Ian
Pearce, Sarah
Mastrano, Alpha
Smith, Rory
Spitler, Lee
Maxted, Nigel
Shabala, Stanislav
Kewley, Lisa
Thrane, Eric
Oslowski, Stefan
Mahony, Elizabeth

Austria

Urbaneja, Miguel
Krásná, Hana
Galiazzo, Mattia
Böhm, Sigrid

Belgium

Triana, Santiago
Mahy Laurent
Tkachenko, Andrii
Defrère, Denis

Brazil

Lima, Eliade
Riffel, Rogério
de Abreu Oliveira, Vinicius
Campos, Fabiola
Martinez, Rafael
Galante, Douglas
Cavichia, Oscar
Selhorst, Caius
Valentim da Costa Lima, Rodolfo
Amôres, Eduardo
WUENSCHE, Carlos
Barres de Almeida, Ulisses

Bulgaria

Boneva, Daniela
Yankova, Krasimira

Canada

Harriott, Tina
Damjanov, Ivana
Lafrenière, David
Liu, Adrian
Majaess, Daniel
Hlozek, Renee
Rowe, Jason
Blumer, Harsha
Sivanandam, Suresh
Kirk, Helen
Rupen, Michael
Cowan, Nicolas
Metchev, Stanimir
Lau, Erwin
Rein, Hanno
Muzzin, Adam
Haggard, Daryl

Chile

Drass, Holger
Jenkins, James
Gunawan, Diah
Treister, Ezequiel
Carrasco, Eleazar
Troncoso Iribarren, Paulina
Salinas, Ricardo
Argudo Fernàndez, Maria del Carmen
Duffau, Sonia
Jaffe, Yara
Ramirez Alegria, Sebastian
Peña Ramírez, Karla
Bayo, Amelia
Alonso-García, Javier
Perez, Laura
Cabrera-Vives, Guillermo

China, Nanjing

HUANG, Maohai
BIAN, Wei-Hao
MENG, Xiangcun
LI, Guoliang
MAO, Jirong
ZHANG, Xiao-Fei
JIANG, Peng
ZHANG, Wen
FANG, Wenjuan
YU, Yun-Wei
FAN, YI-Zhong
LIU, Guilin
WU, Qingwen
CAI, Yi-Fu
YANG, Xiaofeng
WU, jianfeng
WU, Yue
JIANG, Dengkai
XU, Long
YUAN, Haibo
LUO, Jintao
JI, Kaifan
ZHAO, Yinghe
GUO, Wenge

China, Taipei

KOCH, Patrick,
WU, Meng-Ru,
YO, Hwei-Jang,
CHEN, Ke-Jung,
KUO, Cheng-Yu,
CHANG, Chan-Kao,
LIN, Min-Kai,
Lee, Chien-Hsiu

Colombia

Lora Clavijo, Fabio

Czech Republic

Skarka, Skarka
Čadek, Ondřej
Borkar, Abhijeet
Sukovà, Petra
Hanus, Josef
Hornoch, Kamil
Broz, Petr
Hornoch, Kamil
Sebera, Josef
Araudo, Anabella
Jelínek, Petr
Varady, Michal
Dudík, Jaroslav
Zapiór, Maciej

Denmark

Lara-Lopez, Maritza
White, Timothy
Sloth, Martin
Lund, Mikkel

Egypt

Eldaly, Rasha
EMIRA, Mohamed Ghareeb
Abouelmagd, Elbaz
Abdel-Hadi, Yasser
Shaltout, Abdelrazek
Hassan, A. H. Hassan
Awamy, Reda Bendary
Sanad, Magdy
Mohamed, Amaal
semeida, mohamed

Finland

Mursula, Kalevi
Pihajoki, Pauli
Gritsevich, Maria
Mattila, Seppo
Liu, Xiaodong
Veledina, Alexandra
Koljonen, Karri
Abolmasov, Pavel
Janz, Joachim
Markkanen, Johannes

France

Engrand, Cecile
Gavilan, Lisseth
Pitout, Frederic
Langer, Mathieu
Delabrouille, Jacques
Creevey, Orlagh
Allen, Mark
Hennebelle, Patrick
LE VAN SUU, Auguste
Noble, Jennifer
Aulanier, Guillaume
Krim, Lahouari
Thierry, Dudok de Wit
Petrov, Romain
Piétu, Vincent
QUIRICO, Eric
Schüssler, Fabian
Lasue, Jeremie
Fang-Sortais, Bess
Suarez, Olga
Bernard, Edouard
Bacon, Roland
Jean, Pierre
Renaud, Matthieu
Rambaux, Nicolas
Pommier, Mamta
Le Gall, Alice
Blondin, Stéphane
Ziad, Aziz
Bouy, Hervé
Malpangotto, Michela
Brouty, Marianne

Germany

Wiesemeyer, Helmut
Zschocke, Sven
Järvinen, Silva
Irrgang, Andreas
Weingrill, Jörg
Winkel, Benjamin
Weber, Michael
Yeo, Kok Leng
Verma, Meetu
Soler, Juan
Starkenburg, Else
Sasaki, Manami
George, Elizabeth
Adebahr, Björn
Bauswein, Andreas
Lind, Karin
Angelakis, Emmanouil
Castro, Sandra
Shao, Lijing
Aladro Fernandez, Rebeca
Dauser, Thomas
Wieching, Gundolf
Schmidt, Kasper
Cabrera, Juan
Remus, Rhea-Silvia
Fabricius, Maximilian
Gaulme, Patrick
Glawion, Dorit
Nagashima, Kaori
Neumayer, Nadine
Ramsay, Suzanne
Raddi, Roberto
Pidopryhora, Yurii
Ammler-von Eiff, Matthias
Pfalzner, Susanne
Pacaud, Florian
Riquelme Vasquez, Denise
Anderson, Richard
Rottmann, Helge
Wegg, Christopher
Angelou, George
Mugrauer, Markus
Muders, Dirk
Bach, Uwe
Bell, Cameron
Lokanathapura Seetharamabhasari, Anusha
Belloche, Arnaud
Bigiel, Frank
Booth, Mark
Lisker, Thorsten
Bucik, Radoslav
Kamphuis, Peter
Liske, Jochen
Csengeri, Timea
Kruijssen, Diederik
Effenberger, Frederic
Wylezalek, Dominika
Mao, Sui Ann
Fabbian, Damian
Kromer, Markus
Heinkelmann, Robert
Heller, René
Klöckner, Hans-Rainer
Kerzendorf, Wolfgang
Husemann, Bernd
Karuppusamy, Ramesh
Karamanavis, Vassilis (Vasileios)

Greece

Basilakos, Spyros
Bisbas, Thomas
Moraitis, Konstantinos
Antoniadou, Kyriaki I.
Maravelias, Grigoris

Honduras

Mejuto Gonzalez, Javier

Hungary

Szegedi-Elek, Elza
Meszaros, Szabolcs
Regos, Eniko
Plachy, Emese
Kereszturi, Akos
Doherty, Carolyn

India

Sahu, Devendra
Md., Suraj
Jose, Jessy
Kale, Ruta
Misra, Kuntal
Pandey, Biswajit
Pandey, Gajendra
Jain, Rajeev
BHATT, NIPA
Panwar, Neelam
Lekshmi, Resmi
Vig, Sarita
Chattopadhyay, Indranil
Bhattacharyya, Bhaswati
Chattopadhyay, Surajit
Chattopadhyay, Tanuka
CHOWDHURY, PARTHA
Rajaguru, Paul
Banyal, Ravinder
Bangalore, Shylaja
Roy, Jayanta
Kavila, Indulekha
Malu, Siddharth
Datta, Abhirup
Safonova, Margarita
Thakur, Parijat
Ganesh, Shashikiran
Ramanujam, Niruj
AGARAM, RAGHUNATHAN
Saha, Kanak
Aggarwal, Rajiv
Sreeramulu, Sujatha
Vaidya, Bhargav
Kumar, Suresh
Neelamkodan, Naslim
Bagchi, Manjari
Das, Himadri Sekhar
Jain, Deepak
Joshi, Bhuwan
Singh, Veeresh
Bhalerao, Varun
Shah, Priya
Kumar, Nagendra
Roy, Nirupam
Hota, Ananda
Yadav, Vipin
Chandra, Ramesh

Indonesia

Puspitarini, Lucky
Dermawan, Budi
Mumpuni, Emanuel
Aprilia, Aprilia

Ireland

Mackey, Jonathan
McBreen, Sheila
Regan, John
de Mooij, Ernst
Goncharov, Alexander
Fraser, Morgan

Israel

Desjacques, Vincent
Arcavi, Iair
Aharonson, Oded

Italy

Zaharijas, Gabrijela,
Fafone, Viviana
Annibali, Francesca
Israel, Gian Luca
Astone, Pia
Bachetti, Matteo
Zavrtanik, Danilo,
Casu, Silvia
Gruppioni, Carlotta
Vissani, Francesco
Palomba, Cristiano
Christille, Jean Marc
Di Criscienzo, Marcella
Leurini, Silvia
Fiocchi, Mariateresa
Vecchiato, Alberto
Ricci, Fulvio
Fulco, Maria Teresa
Bernardi, Gabriella
Castellano, Marco
Maccone, Claudio

Leaci, Paola
Santini, Paola
Marigo, Paola
Longobardo, Andrea
Yang, Lili,
Vorobiov, Serguei,
Govoni, Federica
Vitale, Stefano
Tomasi, Maurizio

Japan

SALAK, Dragan
FURUYA, Kenji
NAITO, Hiroyuki
NARUSAWA, Shin-ya
USUDA-SATO, Kumiko
EGUCHI, Satoshi
USUI, Fumihiko
KANEKO, Hiroyuki
NAGAMINE, Kentaro
SAWAI, Hidetomo
OHYAMA, Masamitsu
TAKAHASHI, Ryuichi
TORIUMI, Shin
ICHIMOTO, Kiyoshi
HAGIWARA, Yoshiaki
MASADA, Youhei
OOTSUBO, Takafumi
NISHIMURA, Atsushi
WATANABE, Kyoko
MATSUO, Taro
INOUE, Kaiki
MOTOGI, Kazihito
OKOSHI, Katsuya
SHIMONISHI, Takashi
YABUTA, Hikaru
MAEHARA, Hiroyuki
MORIYA, Takashi
UEDA, Akitoshi
AKIYAMA, Eiji
IZUMI, Kiwamu
KARINO, Shigeyuki
MINAMIDANI, Tetsuhiro
IWAI, Kazumasa
OGURI, Masamune
TAKAMI, Kentaro
SANO, Hidetoshi
KAWATE, Tomoko
MATSUOKA, Kenta
SUGAWARA, Yasuharu

Korea, Rep of

SIM, Chae Kyung
HUR, Hyeonoh
KWON, Woojin
AN, DEOKKEUN
JEONG, Il-Gyo
KANG, Sung-Ju
KIM, Mi-Ryang
HOANG, Thiem
LIM, Beomdu
CHUNG, Sun-Ju
CHO, Jungyeon

Lithuania

Maskoliunas, Marius
Stonkutė, Edita
Mikolaitis, Šarūnas
Minkevičiūtė, Renata

Mexico

Alvarez, Cesar
Navarro, Silvana
Gómez Maqueo Chew, Yilen
Nigoche-Netro, Alberto
Rodríguez-González, Ary
Masqué Saumell, Josep Maria
Negrete, Castalia
Franco-Hernández, Ramiro
Delgado Inglada, Gloria
Farah Simón, Alejandro
Torres López, Rosa
Sanchez Sanchez, Sebastian
Rosales-Ortega, Fernando
Carrasco-Gonzalez, Carlos
Márquez Lugo, Ramón Alejandro
Uscanga, Lucero

Netherlands

Broderick, Jess
Birkby, Jayne
Taylor, Matthew
Jennrich, Oliver
Orru, Emanuela
van Weeren, Reinout
Ubachs, Wim
Escoubet, Philippe
Vago, Jorge
Wang, Lingyu
Titov, Dmitrij
Fallows, Richard
Adams, Elizabeth
Bassa, Cornelis
Rowlinson, Antonia
Ferruit, Pierre
Cuppen, Herma
Pizzo, Roberto Francesco
Shimwell, Timothy
Fraternali, Filippo
Guainazzi, Matteo
Offringa, André
Dayal, Pratika
Ormel, Chris

Degenaar, Nathalie
van Bemmel, Ilse
Witasse, Olivier
Mueller, Daniel

New Zealand

Scaringi, Simone
Wiltshire, David
van Straten, Willem

Norway

Shen, Sijing
Tranberg, Anders
Riemer-Sorensen, Signe
Kachelriess, Michael
Wehus, Ingunn Kathrine

Philippines

Deocaris, Custer

Poland

Nalewajko, Krzysztof
Hellwing, Wojciech
Bąk-Stęślicka, Urszula
Hajduk, Marcin
Lewandowski, Wojciech
Kofman, Wlodek
Skowron, Dorota
Mrozek, Tomasz
Kołaczkowski, Zbigniew
Michałowski, Michał
Biesiada, Marek
Mankiewicz, Lech
Oszkiewicz, Dagmara
Dąbrowski, Bartosz
Stawarz, Łukasz
Gromadzki, Mariusz
Orleanski, Piotr
Siarkowski, Marek
Skowron, Jan
Dimitrov, Wojciech
Walczak, Przemysław
Polińska, Magdalena
Pilecki, Bogumił

Portugal

Lima, Raul
Costa, Alexandre
Barros, Susana
Mauricio, Paulo
Adibekyan, Vardan
Buitrago, Fernando

Romania

Besliu-Ionescu, Diana
Ficuţ-Vicaş, Dana
CHIRUTA, Ciprian
Boata, Remus Stefan
Mocanu, Gabriela

Russian Federation

Lapinov, Alexander
Lesovoi, Sergey
Semenko, Eugene
Kudryavtsev, Sergey
Kirsanova, Maria
Sokolova, Marina
Molkov, Sergey
Potekhin, Alexander
Sorokina, Elena
Alekseev, Ilya
Kovalenko, Anatoly
Khaibrakhmanov, Sergey
Barsukova, Elena
Savchenko, Sergey
Vasilyev, Mikhail
Belov, Anatoly
Krivonos, Roman
Vasyunin, Anton
Chernov, Sergey
Kurbatov, Evgeny
Acharova, Irina
Eselevich, Maxim
Melnikov, Aleksander
Galushina, Tatyana
Grankin, Konstantin
Mosenkov, Aleksandr
Kashibadze, Olga
Katkov, Ivan
Pavlov, Dmitry

Serbia, Republic of

Stalevski, Marko

South Africa

Chibueze, James
Józsa, Gyula
Cunnama, Daniel
Frank, Bradley
Pretorius, Magaretha
Evans, Rhodri

Spain

Navarro, Juan F. Navarro
Romero-Gomez, Merce
Aloy, Miguel
Agudo, Ivan
Ebrero, Jacobo
Rodríguez-Gil, Pablo
Bordas, Pol
Armas Padilla, Montserrat
Cardesín Moinelo, Alejandro
Marcos-Arenal, Pablo
Fernández Ontiveros, Juan
García-Rojas, Jorge
González-García, A. César
Mendigutía, Ignacio
Pallé, Pere
Monreal Ibero, Ana
Mora, Alcione

Sweden

Kiss, Mózsi
Finley, Chad
Gudmundsson, Jon
Hallgren, Allan
de la Cruz Rodríguez, Jaime
Mattsson, Lars
Eriksson, Urban
Perez de los Heros, Carlos
Bjerkeli, Per
Marquart, Thomas
Olofsson, Henrik
Stempels, Eric
Church, Ross
Agertz, Oscar
Maercker, Matthias
De Beck, Elvire
Taddia, Francesco
Kainulainen, Jouni
Melinder, Jens
Mustill, Alexander
Adamo, Angela
O'Connor, Evan
Bik, Arjan

Switzerland

Ferrigno, Carlo
Bertone, Stefano
Walter, Benjamin

Thailand

Chantavat, Tooraparb
Chalermsongsak, Tara
Wechakama, Maneenate
Mitthumsiri, Warit
Worrakitpoonpon, Tirawut
Nuntiyakul, Waraporn

Turkey

Yelkenci, Aysegul
Yelkenci, Korhan
NASIROGLU, Ilham
Zengin Çamurdan, Dicle
Topal, Selcuk
OZDEMIR, Sacit
Ozbey Arabaci, Mehtap
Saguner Rambaldi, Tenay
HUDAVERDI, Murat
Benli, Onur
Sezer, Aytap

Ukraine

Shepelev, Valeriy
Panko, Elena
Kovalenko, Nataliya
Akhmetov, Volodymyr
Yankiv-Vitkovska, Liubov
Smerechynskyi, Sviatoslav
Kalinichenko, Mykola
Beshley, Vasyl
Slyusarev, Ivan
Vasylenko, Anatoliy
Dorovskyy, Vladimir
Chinarova, Lidiia
Brazhenko, Anatolii
Breus, Vitalii
Milinevsky, Gennadi
Velichko, Sergey
Apunevych, Stepan

United Kingdom

Jones, Olivia
Keitel, David
Hillier, Andrew
Verscharen, Daniel
Bannister, Michele
Gupta, Jennifer
Cavecchi, Yuri
Mandel, Kaisey
Habergham-Mawson, Stacey
Cole, David
Schoenrich, Ralph
Lewis, Fraser
Jimenez-Serra, Izaskun
Das, Payel
Higgs, Vince
Smith, Keith
Smartt, Stephen
Worley, C. Clare
Jaffe, Andrew
Mortlock, Daniel
Smith, Graham
Woods, Paul
Copperwheat, Christopher
Plane, John
Barstow, Joanna
Kankare, Erkki

United States

Lanz, Lauranne
Hindman, George
Wright, Jason
Nita, Gelu
Law, David
Lang, Cornelia
Wilson-Hodge, Colleen
Hicks, Erin
Santander, Juan
Sargent, Benjamin
Williger, Gerard
Salyk, Colette
Sakari, Charli
Williams, Brian
Whitaker, Katherine
Vilalta, Ricardo
Hartman, Joel
Thyagarajan, Nithyanandan
Vieira, Joaquin
Weinberg, Martin
Vanderbei, Robert
Lai, Dong
Tucker, Douglas
Kolb, Vera
Thorpe, James
Offner, Stella
Rosenfield, Philip
Thompson, Donna
Kohler, Susanna
Hermes, James
SubbaRao, Mark
Oppenheimer, Rebecca
Stuver, Amber
Fischer, William
Cohen, Roger
Stollberg, Mark
Kim, Agnes
Fischer, Debra
Pellerin, Anne
Rice, Emily
Henry, Todd
Finkelstein, Steven
Plazas Malagón, Andrés
Henry, Alaina
Milisavljevic, Danny
Colon, Knicole
Gennaro, Mario
Cooke, William
Perkins, Deborah Kala
Sahnow, David
Correnti, Matteo
Godon, Patrick
Sahlmann, Johannes
Cruz, Kelle
hellbourg, gregory
Cummings, Jeffrey
Henneken, Edwin
McGruder, Charles
Kartaltepe, Jeyhan
Petit, Veronique
Dale, Daniel
Kamionkowski, Marc
Rigby, Jane
Stierwalt, Sabrina
Rieke, George
Jones, Amy
Jin, Meng
Jansen, Rolf
Riechers, Dominik
Ryden, Barbara
Dawson, Rebekah
Momcheva, Ivelina
Privon, George
Richardson, Noel
Pagnotta, Ashley
Salim, Samir
Pisani, Alice
Roman-Duval, Julia
James, Bethan
Danehkar, Ashkbiz
Stevenson, Kevin
Daou, Doris
Frye, Brenda
Jacobson, Seth
Desjardins, Tyler
Raab, Frederick
Robinson, Frank
Reitzel, David
Indriolo, Nicholas
Rangelov, Blagoy
Durst, Steve
Ragozzine, Darin
Hummels, Cameron
Nair, Preethi
Hines, Dean
Fan, Xiaohui
Steinhauer, Aaron
Rieke, Marcia
Riggs, Jamie
Freeman, Peter
Pankratius, Victor
Pesce, Joseph
Li, Pak Shing
Meyer, Eileen
SHUCH, H. Paul
Zemcov, Michael
Monroe, TalaWanda
Sterling, Nicholas
Milam, Stefanie
Natarajan, Priyamvada
Nugent, Carolyn

Lindner, Rudi
Moorhead, Althea
Moran, Christine
Cavaglia, Marco
Zabludoff, Ann
Murphy, Eric
Haqq-Misra, Jacob
Siemion, Andrew
Wong, Ka-Wah
Gaspari, Massimo
Meinke, Bonnie
Frouard, Julien
McLaughlin, Maura
Nataf, David
Shirley, Yancy
McGuire, Brett
Dupuy, Trent
Diaz, Rosa
Marshall, Jennifer
O'Shaughnessy, Richard
LaMassa, Stephanie
McQuinn, Kristen
Whelan, John
Neilsen, Eric
Olenick, Richard
Sohn, Sangmo
Genel, Shy
Sokoloski, Jennifer (Jeno)
Marcucci, Emma
Lee, Nicolas
Welty, Daniel
Ly, Chun
Bottorff, Mark
Lugaz, Noé
Close, Sigrid
Gawiser, Eric
Lucas, Ray
Noel-Storr, Jacob
Way, Michael
Chizek Frouard, Malynda
Lynch, Ryan
Gorski, Krzysztof
Chen, Christine
MacMillan, Daniel
van der Horst, Alexander
Gosnell, Natalie
Chen, Bin
Weryk, Robert
Casey, Caitlin
Lubow, Stephen
Carlberg, Joleen
Garrod, Robin
Bryan, Greg
Bruursema, Justice
Brown, Peter
Boyer, Martha
Loredo, Thomas
Tollerud, Erik
Kannappan, Sheila
Borthakur, Sanchayeeta
Sun, Xudong
Bhatnagar, Sanjay
Berghea, Ciprian
somerville, rachel
Rubin, David
Su, Kate
Berger, Thomas
Battersby, Cara
Barentine, John
De Rosa, Gisella
GAi, Moshe
Giaime, Joseph
Barclay, Thomas
PAN, Kuo-Chuan
del Pino Molina, Andres
Arcand, Kimberly
Araya, Esteban
Avila, Roberto
Besla, Gurtina
Friesen, Rachel
Annis, James
French, Rica
Scott, Nicholas
Alpaslan, Mehmet
Siemiginowska, Aneta
Zweibel, Ellen
Allen, Alice
Sewilo, Marta
Liu, Charles
Barentsen, Geert
Loebman, Sarah
Lister, Timothy
Schwamb, Megan
Li, Yuan
Leutenegger, Maurice
Armus, Lee
Allam, Sahar
Adhikari, Rana
Collinge, Matt
Merényi, Erzsébet

Vatican City State

D Souza, Richard
Galaverni, Matteo
Gionti, Gabriele

DECEASED MEMBERS (2015–2018) REMEMBERED

Abalakin, Viktor
Abulazm, Mohamed
A'Hearn, Michael
Akimov, Leonid
Alexander, Tal
Alexandrov, Yuri
Ali, Gamal
Alksnis, Andrejs
Alley, Carrol
Altenhoff, Wilhelm
Argon, Alice
Aslanov, I.
Aurass, Henry
Barkin, Yuri
Bar-Nun, Akiva
Bekenstein, Jacob
Bell, Barbara
Bell, Kenneth

Belton, Michael
Benevides Soares, Paulo
Bhatnagar, K.
Bignami, Giovanni
Binetruy, Pierre
Bless, Robert
Bohm-Vitense, Erika
Bonnor, W.
Borisova, Ana
Boyarchuk, Alexander
Brahic, André
Brownlee, Robert
Bruns, Andrey
Bumba, Vaclav
Cameron, Winifred
Castets, Alain
CHEN, Yafeng
Chertoprud, Vadim
Chupp, Edward
Churyumov, Klim
Coleman, Paul
Colín, Pedro
Combes, Michel
Crotts, Arlin
Cuypers, Jan
Dadaev, Aleksandr
Davies, Rodney
Davis, Richard
de Bruyn, A.
Delaboudinère, Jean-Pierre
Delsemme, Armand
Deubner, Franz-Ludwig
Divan, Lucienne
Dolidze, Madona
Donner, Karl
Doroshenko, Valentina
Dudinov, Volodymyr
Dumont, René
Dzhapiashvili, Victor
Dzigvashvili, R.
Elford, William
Elste, Günther
Elston, Wolfgang
Eminzade, T.
Erickson, William
Eshleman, Von
Ewen, Harold
Florsch, Alphonse
Fong, Richard
Freire Ferrero, Rubens
Fursenko, Margarita
Galal, A.
García-Berro, Enrique
Garrison, Robert
Gathier, Roel
Gebbie, Katharine
Gehrels, Neil
Giannone, Pietro
Gnedin, Yurij
Golovatyj, Volodymyr
Gorschkov, Aleksandr
Gosachinskij, Igor
Goudas, Constantine
Greyber, Howard
Grib, Sergei
Guinot, Bernard
Gusejnov, Ragim
Hady, Ahmed
Halliday, Ian
Hamdy, M.
Harris, Daniel
HASEGAWA, Ichiro
Hauck, Bernard
Haupt, Hermann
Hawking, Stephen
Heddle, Douglas
Hide, Raymond
Hotinli, Metin
Houck, James
Howard III, William
Hurnik, Hieronim
HURUKAWA, Kiichirō
Hysom, Edmund
ISHIMARU, Yuhri
Ivanov-Kholodny, Gor
Jahn, Krzysztof
Jiang, Yunchun
Jura, Michael
Kalenichenko, Valentin
Kasumov, Fikret
Kaufmann, Pierre
Keay, Colin
Kebede, Legesse
Kendziorra, Eckhard
Kołaczek, Barbara
Komberg, Boris
KONDO, Yoji
Kononovich, Edvard
Korzhavin, Anatoly
Kovalevsky, Jean
KOZAI, Yoshihide
Kraft, Robert
Kroto, Harold
Kubičela, Aleksandar
Kuimov, Konstantin
Landi Degl'Innocenti, Egidio
Lane, Adair
Lawrence, John
LI, Guangyu
Linnell, Albert
Livshits, Moisey
Lo, Fred K. Y.
Loden, Lars
Lowe, Robert
LU, Jufu
Lucy, Leon
Lynden-Bell, Donald
Magnan, Christian
Malaroda, Stella
Malyuto, Valeri
Marie, Mohamed
Markkanen, Tapio
McClintock, Jeffrey
McIntosh, Patrick
Meadows, A.
Meire, Raphael
Mestel, Leon
Michel, F.
Mihajlov, Anatolij
Miletsky, Eugeny
Mitalas, Romas
MIZUTANI, Kohei
Moehlmann, Diedrich
Murray, Stephen
NAKAYAMA, Kunji
NAN, Ren-Dong
Nelson, Jerry
Oblak, Edouard
Østgaard, Erlend
Olowin, Ronald
Orlov, Viktor
OWAKI, Naoaki
Owen, Tobias
Palla, Francesco
Palmer, Philip
Palumbo, Giorgio
PARK, Hong-Seo
Paternò, Lucio
Pauls, Thomas
Petrov, G.
Petrova, Svetlana
Pflug, Klaus

Philip, A.G.
Pittich, Eduard
Polcaro, V.
Polosukhina-Chuvaeva, Nina
Pottasch, Stuart
Pronik, Vladimir
Revnivtsev, Mikhail
Richards, Mercedes
Roemer, Elizabeth
Roennaeng, Bernt
Röser, Hans-peter
Rohlfs, Kristen
Romanov, Yuri
Rubin, Vera
Ruder, Hanns
Ruskol, Evgeniya
Rusu, I.
Sattarov, Isroil
Savage, Ann
Schmadel, Lutz
Schmidt-Kaler, Theodor
Scholz, Michael
Schutz, Bob
Semeniuk, Irena
Shakeshaft, John
Shaltout, Mosalam
Shishov, Vladimir
Siregar, Suryadi
Slee, O.
Smith, Bradford
Sorochenko, Roman
Spielfiedel, Annie
Spinrad, Hyron
Stecher, Theodore
Steigman, Gary
Steinberg, Jean-Louis
Steinlin, Uli
Steshenko, N.
Sultanov, G.
Swenson Jr., George
Tanaka, Yasuo
Taranova, Olga
Tassoul, Monique
Teplitskaya, Raisa
Teske, Richard
Thaddeus, Patrick
Tiuri, Martti
Tofani, Gianni
Trefftz, Eleonore
Upgren, Arthur
Valyaev, Valerij
Vityazev, Veniamin
Voigt, Hans
Voshchinnikov, Nikolai
Walborn, Nolan
Walker, Helen
Wallace, Lloyd
Walmsley, C.
WANG, Zhenru
Weaver, Harold
Weiler, Kurt
Wesson, Paul
Whitaker, Ewen
Whitney, Charles
Winiarski, Maciej
Winkler, Gernot
Wiramihardja, Suhardja
Wolstencroft, Ramon
Yang, Zhiliang
Yanovitskij, Edgard
Young, Louise
Yuldashbaev, Taimas
Zander, Rodolphe
Zhang, Yuying
ZHANG, Jialu
Zombeck, Martin

Transactions IAU, Volume XXX
Proc. XXX IAU General Assembly, August 2018
Debra Elmegreen, ed.
doi:10.1017/S1743921323005689

CHAPTER X

THIRTIETH GENERAL ASSEMBLY

Divisions, Commissions and Working Groups

for the Triennium 2015–2018

This chapter lists the main IAU scientific bodies: nine Divisions plus their associated Commissions and Working Groups. In addition to these, there are several Executive Committee Working Groups that are also presented.

Division A Fundamental Astronomy

President:
Anne Lemaître
University of Namur
Namur Center of Complex Systems (naXys)
Rempart de la Vierge 8
5000–Namur
Belgium

Vice-President:
Daniel Hestroffer
Observatoire de Paris
IMCCE
77 Av Denfert Rochereau–Bat. A
75014–Paris
France

Secretary:
Susan Gessner Stewart
United States Naval Observatory
Astronomical Applications
3450 Massachusetts Ave., NW
20392-5420 Washington
United States

Steering Committee Members:

Elisa Felicitas Arias (France), Fernando Virgilio Roig (Brazil), Cristian Beaugé (Argentina), William M. Folkner (United States), Anne Lemaître (Belgium), Sergei A. Klioner (Germany), Anthony G.A. Brown (Netherlands), Richard Gross (United States), Susan Gessner Stewart (United States), Nicole Capitaine (France), Catherine Y. Hohenkerk (United Kingdom), Daniel Hestroffer (France), Sylvio Ferraz-Mello (Brazil), Ralph A. Gaume (United States)

Commissions:

Commission A1 Astrometry
Commission A2 Rotation of the Earth
Commission A3 Fundamental Standards
Inter-Division A-F Commission Celestial Mechanics and Dynamical Astronomy
Cross-Division A-F Commission Solar System Ephemerides

Division B Facilities, Technologies and Data Science

President:
Pietro Ubertini
INAF
IAPS–Institute for Space Astrophysics and Planetology
Via Fosso del Cavaliere 100
00133–Roma
Italy

Vice-President:
Michael G. Burton
University of New South Wales
School of Physics
NSW 2052–Sydney
Australia

Secretary:
Michael G. Burton
University of New South Wales
School of Physics
NSW 2052–Sydney
Australia

Steering Committee Members:

Eric D. Feigelson (United States), Ana I. Gómez de Castro (Spain), Peter Joseph Quinn (Australia), Wenwu TIAN (China Nanjing), Gabriele Giovannini (Italy), Farid Salama (United States), Saul J. Adelman (United States), David Richard Silva (United States), Pietro Ubertini (Italy), Michael G. Burton (Australia), Simon F. Portegies Zwart (Netherlands), Richard F. Green (United States), Lisa Storrie-Lombardi (United States), Michael W. Wise (Netherlands)

Commissions:

Commission B1 Computational Astrophysics
Commission B2 Data and Documentation
Commission B3 Astroinformatics and Astrostatistics
Commission B4 Radio Astronomy
Commission B5 Laboratory Astrophysics
Commission B6 Astronomical Photometry and Polarimetry
Inter-Division B-C Commission Protection of Existing and Potential Observatory Sites
Inter-Division B-H-J Commission Intergalactic Medium

Division C Education, Outreach and Heritage

President:
John B. Hearnshaw
University of Canterbury
Physics-Astronomy Department
Private Bag 4800
8020–Christchurch
New Zealand

Vice-President:
Susana E. Deustua
Space Telescope Science Inst
Instruments Division
3700 San Martin Dr
MD 21218–Baltimore
United States

Secretary:
Pamela L. Gay
Southern Ilinois University Edwardsville
Center for STEM Research, Education, & Outreach
Center for STEM REO
SIUE, Campus Box 2224
62025–Edwardsville
United States

Steering Committee Members:

Katrien Kolenberg (Belgium), Saeko S. Hayashi (United States), Mary Kay M. Hemenway (United States), John B. Hearnshaw (New Zealand), Susana E. Deustua (United States), Pamela L. Gay (United States), Beatriz Elena García (Argentina), Pedro Russo (Netherlands), Xiaochun Sun (China Nanjing), Clive Ruggles (United Kingdom), Michele Gerbaldi (France), Kazuhiro SEKIGUCHI (Japan), Linda Elisabeth Strubbe (Canada)

Commissions:

Inter-Division B-C Commission Protection of Existing and Potential Observatory Sites
Commission C1 Astronomy Education and Development
Commission C2 Communicating Astronomy with the Public
Commission C3 History of Astronomy
Commission C4 World Heritage and Astronomy

Division D High Energy Phenomena and Fundamental Physics

President:
Chryssa Kouveliotou
George Washington University
Physics Department
Corcoran Hall
725 21st Street, NW
DC 20052–Washington
United States

Vice-President:
Elena Pian
INAF
Institute of Space Astrophysics and Cosmic Physics
Via P. Gobetti 101
40129–Bologna
Italy

Secretary:
Anna L Watts
University of Amsterdam
Astronomical Institute 'Anton Pannekoek'
Postbus 94249
1090 GE–Amsterdam
Netherlands

Steering Committee Members:

Anna L Watts (Netherlands), Tadayasu DOTANI (Japan), Isabelle Grenier (France), Neil Gehrels (United States), Diana Mary Worrall (United Kingdom), Chryssa Kouveliotou (United States), William Richard Forman (United States), Xavier Barcons (Spain), John G. Kirk (Germany), Anna Wolter (Italy), Elena Pian (Italy)

Commissions:

Commission D1 Gravitational Wave Astrophysics
Inter-Division D-G-H-J Commission Galaxy Spectral Energy Distributions
Cross-Division D-J Commission Supermassive Black Holes, Feedback and Galaxy Evolution

Division E Sun and Heliosphere

President:
Yihua Yan
National Astronomical Observatories
20A Datun Rd
Chaoyang District
100012 - Beijing
China Nanjing

Vice-President:
Sarah Gibson
NCAR/HAO
p.o. box 3000
80301-3000–Boulder
United States

Secretary:
Marc L. DeRosa
Lockheed Martin Solar and Astrophysics Laboratory
3251 Hanover St. B/252
94304–Palo Alto
United States

Steering Committee Members:

Toshifumi Shimizu (Japan), Eduard P. Kontar (United Kingdom), Nandita Srivastava (India), Natalie Krivova (Germany), Lidia van Driel-Gesztelyi (United Kingdom), Yihua YAN (China Nanjing), Lyndsay Fletcher (United Kingdom), Sarah Gibson (United States), Ingrid Mann (Sweden), Marc L DeRosa (United States), Arnab Rai Choudhuri (India), Rudolf von Steiger (Switzerland)

Commissions:

Commission E1 Solar Radiation and Structure
Commission E2 Solar Activity
Commission E3 Solar Impact Throughout the Heliosphere

Division F Planetary Systems and Bioastronomy

President:
Nader Haghighipour
University of Hawaii
Institute for Astronomy
2680 Woodlawn Dr
HI 96822 - Honolulu
United States

Vice-President:
Gonzalo Tancredi
Facultad de Ciencias
Departamento de Astronomía
Igua 4225
11400–Montevideo
Uruguay

Secretary:
Jean-Luc Margot
University of California
595 Charles E Young Drive East
CA 90095 –Los Angeles
United States

Steering Committee Members:

Maria A. Barucci (France), Sun Kwok (China Nanjing), Daniela Lazzaro (Brazil), Andrea Milani Comparetti (Italy), Nader Haghighipour (United States), Gonzalo Tancredi (Uruguay), Giovanni B. Valsecchi (Italy), Jiří Borovička (Czech Republic), Alain Lecavelier des Etangs (France), Jean-Luc Margot (United States), Paul Winchester Chodas (United States), Athena Coustenis (France), William M. Irvine (United States), Didier Queloz (United Kingdom)

Commissions:

Inter-Division A-F Commission Celestial Mechanics and Dynamical Astronomy
Commission F1 Meteors, Meteorites and Interplanetary Dust
Commission F2 Exoplanets and the Solar system
Commission F3 Astrobiology
Cross-Division A-F Commission Solar System Ephemerides

Division G Stars and Stellar Physics

President:
Corinne Charbonnel
Geneva Observatory (Switzerland)
& CNRS (France)
Chemin des Maillettes 51
1290–Versoix
Switzerland

Vice-President:
David R. Soderblom
Space Telescope Science Institute
3700 San Martin Dr
MD 21218-2410–Baltimore
United States

Secretary:
Tabetha Suzanne Boyajian
Louisiana State University
Department of Physics and Astronomy
Baton Rouge
United States

Steering Committee Members:

Francesca D'Antona (Italy), Tabetha Suzanne Boyajian (United States), Andrej Prša (United States), Artemio Herrero Davó (Spain), John C. Lattanzio (Australia), Christopher Simon Jeffery (United Kingdom), Ivan Hubeny (United States), Pierre Kervella (France), Geraldine Joan Peters (United States), Corinne Charbonnel (Switzerland), David R. Soderblom (United States), Ignasi Ribas (Spain), Martin Asplund (Australia)

Commissions:

Commission G1 Binary and Multiple Star Systems
Commission G2 Massive Stars
Commission G3 Stellar Evolution
Commission G4 Pulsating Stars
Commission G5 Stellar and Planetary Atmospheres
Inter-Division G-H-J Commission Stellar Clusters throughout Cosmic Space and Time
Inter-Division D-G-H-J Commission Galaxy Spectral Energy Distributions

Division H Interstellar Matter and Local Universe

President:
Bruce G. Elmegreen
IBM
T J Watson Res Center
1101 Kitchawan Road
10598–Yorktown Heights
United States

Vice-President:
Leonardo Testi
INAF
Osservatorio Astrofisico di Arcetri
Largo E Fermi 5
50125–Firenze
Italy

Secretary:
Francisca Kemper
Academia Sinica
Institute of Astronomy and Astrophysics
PO Box 23-141 10617 - Taipei
China Taipei

Organizing Committee Members:

Holger Baumgardt (Australia), Eva Schinnerer (Germany), Cristina Chiappini (Germany), Eva K. Grebel (Germany), Thomas J. Millar (United Kingdom), Letizia Stanghellini (United States), Richard de Grijs (China Nanjing), Bruce G. Elmegreen

(United States), Leonardo Testi (Italy), Francisca Kemper (China Taipei), Ewine F. van Dishoeck (Netherlands), Diego Mardones (Chile), Michael R. Meyer (Switzerland)

Commissions:

Commission H1 The Local Universe
Commission H2 Astrochemistry
Commission H3 Planetary Nebulae
Inter-Division G-H-J Commission Stellar Clusters throughout Cosmic Space and Time
Inter-Division D-G-H-J Commission Galaxy Spectral Energy Distributions
Inter-Division B-H-J Commission Intergalactic Medium

Division J Galaxies and Cosmology

President:
Claus Leitherer
Space Telescope Science Inst
3700 San Martin Dr
MD 21218-2410–Baltimore
United States

Vice-President:
Matthew Arnold Malkan
UCLA
Physics-Astronomy Dept
Box 951547
405 Hilgard Ave
CA 90095-1547–Los Angeles
United States

Secretary:
Jeremy R. Mould
Swinburne University
Centre for Astrophysics and Supercomputing
John St
Hawthorn–3122
Melbourne
Australia

Organizing Committee Members:
Andrew J. Bunker (United Kingdom), Monica Rubio (Chile), Leslie Hunt (Italy), Marcella Carollo (Switzerland), Jeremy R. Mould (Australia), Denis Burgarella (France), Avery A. Meiksin (United Kingdom), Claus Leitherer (United States), Matthew Arnold Malkan (United States), Françoise Combes (France), Thaisa Storchi-Bergmann (Brazil), Stéphane J. Courteau (Canada)

Commissions:

Inter-Division G-H-J Commission Stellar Clusters throughout Cosmic Space and Time
Inter-Division D-G-H-J Commission Galaxy Spectral Energy Distributions
Inter-Division B-H-J Commission Intergalactic Medium
Cross-Division D-J Commission Supermassive Black Holes, Feedback and Galaxy Evolution

Division Working Groups

Division	*Working Group Name*	*Chair*	*Co-Chair(s)*
D A	Astrometry by Small Ground-Based Telescopes		William Thuillot, Marcelo Assafin
D A	Multi-waveband Realisations of International Celestial Reference System	François Mignard	
D A	Numerical Standards in Fundamental Astronomy (NSFA)	Brian Luzum	
D A	Standards of Fundamental Astronomy (SOFA)	C. Y. Hohenkerk	
D A	Third Realisation of International Celestial Reference Frame	Patrick Charlot	
D B	Time Domain Astronomy	Rob Seaman	
D B	UV Astronomy	A. I. Gómez de Castro	
D C	Star Names	Eric E. Mamajek	
D D	Supernovae	Avishay Gal-Yam	
D E	Impact of Magnetic Activity on Solar and Stellar Environments	Dibyendu Nandi	
D F	Exoplanets for the Public	A. Lecavelier des Etangs	
D F	Near Earth Objects	Patrick Michel	
D F	Planetary System Nomenclature (WGPSN)	Rita M. Schulz	
D F	Small Bodies Nomenclature (SBN)	Jana Tichá	
D G	Abundances in Red Giants	Jacco Th. van Loon	
D G	Active B Stars	Carol Evelyn Jones	
D G	Ap & Related Stars	Gautier Mathys	
D H	The Galactic Center		

Inter-Division Working Groups

Division	*Working Group Name*	*Chair*	*Co-Chair(s)*
D A-F	Cartographic Coordinates & Rotational Elements	Brent A. Archinal	
D B E	Coordination of Synoptic Observations of the Sun		Frederic Clette, Alexei A. Pevtsov
D C-E	Solar Eclipses	Jay M. Pasachoff	

Commission Working Groups

Commission	*Working Group Name*	*Chair*	*Co-Chair(s)*
C A2	Theory of Earth Rotation and Validation (IAU / IAG Joint WG)	Jose M. Ferrándiz	
C B4	Historic Radio Astronomy	R. Wielebinski	R. T. Schilizzi
C B5	High-Accuracy Stellar Spectroscopy	Paul S. Barklem	
C B7	Site Protection	Richard F. Green	Saeko S. Hayashi
C B7	Technical Working Group	Diane E. Turnshek	
C C1	Astronomy for Equity and Inclusion	Amelia Ortiz Gil	
C C1	Network for Astronomy School Education (NASE)	Rosa M. Ros	Beatriz E. García
C C1	Theory and Methods in Astronomy Education	Paulo S. Bretones	
C C2	CAP Conferences	Sze-leung Cheung	Oana M. Sandu
C C2	CAP Journal	Georgia R. Bladon	

Commission	*Working Group Name*	*Chair*	*Co-Chair(s)*
C C2	Outreach Professionalization & Accreditation	Richard T. Fienberg	
C C2	Public Outreach Information Management	Lars L. Christensen	
C C2	Science Communication Research in Astronomy	Marta Entradas	
C C3	Historical Radio Astronomy (WGHRA)		
C C3	Johannes Kepler	Terence J. Mahoney	
C C4	Astronomical Heritage in Danger	Alejandro M. López	
C C4	Classical Observatories from the Renaissance to the 20th Century	Gudrun B. E. Wolfschmidt	
C C4	Heritage of Space Exploration	Mikhail Ya Marov	
C E1	Solar Irradiance	Greg Kopp	A. I. Shapiro
C F1	Meteor Shower Nomenclature (MSN-WG)	Diego Janches	

Inter-Commission Working Groups

Commission	*Working Group Name*	*Chair*
CB7,CC1	Achieving Sustainable Development within a Quality Lighting Framework	
CB7,CC4	Windows to the Universe: High-Mountain Observatories, and Other Astronomical Sites of the Late 20th and Early 21st Centuries	Robert C. Smith
CC1,CF2,CF3,CH2	Education and Training in Astrobiology	Muriel Gargaud
CC1,CC4	Intangible Astronomical Heritage	Duane W. Hamacher
CC3,CC4	Archaeoastronomy and Astronomy in Culture	Raymond P. Norris

Executive Committee Working Groups

Working Group Name	*Chair*	*Co-Chair(s)*
Dark and Quiet Sky Protection	Richard F. Green	
Global Coordination of Ground and Space Astrophysics	Roger L. Davies	David N. Spergel
International Year of Light: Cosmic Light	Richard F. Green	
Public Naming of Planets and Planetary Satellites	Thierry Montmerle	
Women in Astronomy	Francesca Primas	

Commissions (2015–2018)

Commission A1–Astrometry

President:
Anthony G.A. Brown
Leiden University
Leiden Observatory
PO Box 9513
2300 RA–Leiden
Netherlands

Vice-President:
Jean Souchay
Observatoire de Paris
SYRTE
61 Av de l'Observatoire
75014–Paris
France

Secretary:
Dafydd Wyn Evans
University of Cambridge
Institute of Astronomy
Madingley Road
CB3 0HA–Cambridge
United Kingdom

Organizing Committee Members:
Jean Souchay (France), Dafydd Wyn Evans (United Kingdom), Alexandre Humberto Andrei (Brazil), Stephen C. Unwin (United States), Yoshiyuki YAMADA (Japan), Norbert Zacharias (United States), Anthony G.A. Brown (Netherlands)

Commission A2–Rotation of the Earth

President:
Richard Gross
Jet Propulsion Laboratory
Geodynamics and Space Geodesy Group
Mail Stop 238-600
4800 Oak Grove Drive
91109–Pasadena, CA
United States

Vice-President:
Florian Seitz
Technische Universität München
Deutsches Geodätisches Forschungsinstitut (DGFI)
Arcisstr. 21
D-80333–Munich
Germany

Secretary:
Alberto Escapa
Universidad de Alicante
Matemática Aplicada. EPS
Campus de San Vicente
03690–San Vicente de Raspeig–Alicante
Spain

Organizing Committee Members:
Jose Manuel Ferrándiz (Spain), Florian Seitz (Germany), Alberto Escapa (Spain), Daniela Thaller (Germany), Vladimir E. Zharov (Russian Federation), Richard Gross (United States)

Commission A3–Fundamental Standards

President:
Catherine Y. Hohenkerk
UKHO
HM Nautical Almanac Office
Admiralty Way
TA1 2DN–Taunton
United Kingdom

Vice-President:
Brian Luzum
US Naval Observatory
3450 Massachusetts Ave NW
DC 20392 5420–Washington
United States

Secretary:
Nancy A. Oliversen
United States Naval Observatory
Astronomical Applications Department
3450 Massachusetts Ave, N. W.
20392-5410–Washington
United States

Organizing Committee Members:

Nicole Capitaine (France), Catherine Y. Hohenkerk (United Kingdom), Brian Luzum (United States), Charles H. Acton (United States), John A. Bangert (United States), Dennis D. McCarthy (United States), Nancy A. Oliversen (United States)

Commission B1–Computational Astrophysics

President:
Simon F. Portegies Zwart
Leiden Observatory
Leiden University
PO Box 9513
NL2300 RA–Leiden
Netherlands

Vice-President:
Dmitrij V. Bisikalo
Inst of Astronomy RAS
Pyatnitskaya ul 48
119017–Moscow
Russian Federation

Secretary:
Christian M. Boily
Observatoire Astronomique de Strasbourg
11 rue de l'Université
67000–Strasbourg
France

Organizing Committee Members:

Dmitrij V. Bisikalo (Russian Federation), Sungsoo S. KIM (Korea, Rep of), Irina N. Kitiashvili (United States), Garrelt Mellema (Sweden), Michael Shara (United States), Christian M. Boily (France), Simon F. Portegies Zwart (Netherlands)

Commission B2–Data and Documentation

President:
Michael W. Wise
ASTRON
Netherlands Institute for Radio Astronomy
P.O. Box 2
7990 AA–Dwingeloo
Netherlands

Vice-President:
Anja C. Schröder
SAAO
Observatory Road
P O Box 9
7935–Cape Town
South Africa

Organizing Committee Members:

Michael W. Wise (Netherlands), Anja C. Schröder (South Africa), Chenzhou CUI (China Nanjing), R. Elizabeth M. Griffin (Canada), Robert J. Hanisch (United States), Arnold H. Rots (United States), Rob Seaman (United States)

Commission B3–Astroinformatics and Astrostatistics

President:
Eric D. Feigelson
Pennsylvania State University
Astronomy & Astrophysics Dept
525 Davey Lab
PA 16802–University Park
United States

Vice-President:
Prajval Shastri
Indian Institute of Astrophysics
Sarjapur Road
Koramangala
560 034–Bengaluru
India

Secretary:
Saeqa Dil Vrtilek
Harvard Smithsonian CfA
HCO/SAO
MS 67
60 Garden Str
MA 02138-1516–Cambridge
United States

Organizing Committee Members:

Eric D. Feigelson (United States), Saeqa Dil Vrtilek (United States), Eric B. Ford (United States), Alan Heavens (United Kingdom), Fionn Murtagh (United Kingdom), Yanxia ZHANG (China Nanjing), Prajval Shastri (India)

Commission B4–Radio Astronomy

President:
Gabriele Giovannini
INAF
Istituto di Radioastronomia
Via P Gobetti 101
40129–Bologna
Italy

Vice-President:
Anthony James Beasley
NRAO
520 Edgemont Rd
22903-2475–Charlottesville
United States

Secretary:
Wim van Driel
Observatoire Paris-Site de Meudon
GEPI
5 Pl Jules Janssen
92195–Meudon Cedex
France

Organizing Committee Members:

Anthony James Beasley (United States), Gabriele Giovannini (Italy), Wim van Driel (France), Xiaoyu HONG (China Nanjing), Joseph Lazio (United States), Nicholas Seymour (Australia), Jan Mathijs van der Hulst (Netherlands), Tony H. Wong (United States)

Commission B5–Laboratory Astrophysics

President:
Farid Salama
NASA - Ames Research Center
Space Science & Astrobiology Division
Building N245
MS 245 6
CA 94035-1000–Moffett Field
United States

Vice-President:
Helen J. Fraser
The Open Unviersity
Department of Physical Sciences,
Astronomy Division
Walton Hall
MK7 6AA–Milton Keynes
United Kingdom

Secretary:
Gianfranco Vidali
Syracuse University
Department of Physics
221 Physics Building
NY 13244-1130–Syracuse
United States

Organizing Committee Members:

Paul S. Barklem (Sweden), Farid Salama (United States), Gianfranco Vidali (United States), Thomas Henning (Germany), Harold Linnartz (Netherlands), Feilu WANG (China Nanjing), Helen J. Fraser (United Kingdom)

Commission B6–Astronomical Photometry and Polarimetry

President:
Saul J. Adelman
The Citadel
Physics Dept
171 Moultrie St
SC 29409-0270–Charleston
United States

Vice-President:
Antonio Mario Magalhaes
Universidade de São Paulo
Instituto de Astronomia, Geofisica e Ciencias Atmosfericas
Rua do Matão 1226
Cidade Universitaria
05508-090–São Paulo
Brazil

Secretary:
Saul J. Adelman
The Citadel
Physics Dept
171 Moultrie St

SC 29409-0270–Charleston
United States

Organizing Committee Members:

Antonio Mario Magalhaes (Brazil), Saul J. Adelman (United States), Pierre Bastien (Canada), Richard Ignace (United States), Carme Jordi (Spain), J. Allyn Smith (United States), Kevin Volk (United States)

Commission C1–Astronomy Education and Development

President:
Beatriz Elena García
Instituto de Tecnologías en Detección y Astropartículas (ITeDA)
ITeDA Mendoza
Azopardo 313
Godoy Cruz
5501–Mendoza
Argentina

Vice-President:
Paulo Sergio Bretones
Universidade Federal de São Carlos
Departamento de Metodologia de Ensino
13565-905–São Carlos
Brazil

Secretary:
Kathleen DeGioia Eastwood
Northern Arizona University
Physics and Astronomy Department
Box 6010
AZ 86011-6010–Flagstaff
United States

Organizing Committee Members:

Beatriz Elena García (Argentina), Paulo Sergio Bretones (Brazil), Jean-Pierre de Greve (Belgium), Christopher David Impey (United States), Nicoletta Lanciano (Italy), Amelia Ortiz Gil (Spain), Kathleen DeGioia Eastwood (United States)

Commission C2–Communicating Astronomy with the Public

President:
Pedro Russo
Leiden Observatory
Universe Awareness International Office
P.O. Box 9513
NL-2300 RA–Leiden
Netherlands

Vice-President:
Richard Tresch Fienberg
American Astronomical Society
43 Everett Ave.
MA 02472-1881–Watertown
United States

Secretary:
Kingsley C. Okpala
University of Nigeria
Physics & Astronomy
University Road, Nsukka, Enugu State, Nigeria
41001–Nsukka, Enugu State
Nigeria

Organizing Committee Members:

Pedro Russo (Netherlands), Kingsley C. Okpala (Nigeria), Megan Kirsty Argo (United Kingdom), Lars Lindberg Christensen (Germany), Carol Ann Christian (United States), Sylvie D. Vauclair (France), Richard Tresch Fienberg (United States)

Commission C3–History of Astronomy

President:
Xiaochun Sun
Inst History Natural Science, CAS
55 Zhongguancun East Road
100190–Beijing
China Nanjing

Vice-President:
Wayne Orchiston
National Astronomical Research Institute of Thailand
191 Huay Kaew Road
Suthep District
50200–Chiang Mai
Thailand

Organizing Committee Members:

Xiaochun Sun (China Nanjing), Wayne Orchiston (Thailand), Owen Gingerich (United States), Raymond P. Norris (Australia), Jay M. Pasachoff (United States), Christiaan L. Sterken (Belgium), David Valls-Gabaud (France)

Commission C4–World Heritage and Astronomy

President:
Clive Ruggles
University of Leicester
School of Archaeology and Ancient History
University Road
LE1 7RH–Leicester
United Kingdom

Vice-President:
Gudrun B. E. Wolfschmidt
Hamburg University, Center for history of science and technology
Department of Physics, Hamburg Observatory
Bundesstr 55 Geomatikum
D-20146–Hamburg
Germany

Secretary:
Roger Ferlet
UMR7095 CNRS
Institut d'Astrophysique de Paris
98bis Bd Arago
75014–Paris
France

Organizing Committee Members:

Roger Ferlet (France), Clive Ruggles (United Kingdom), Gudrun B. E. Wolfschmidt (Germany), Siramas Komonjinda (Thailand), Mikhail Ya Marov (Russian Federation), Malcolm G. Smith (Chile)

Commission D1–Gravitational Wave Astrophysics

President:
Neil Gehrels
NASA/GSFC
Code 661
MD 20771–Greenbelt
United States

Vice-President:
Marica Branchesi
Università degli studi di Urbino
"Carlo Bo"
Dipartimento di Scienze di Base e
Fondamenti/Physics section
Via Santa Chiara, 27
61029–Urbino
Italy

Organizing Committee Members:

Neil Gehrels (United States), Marica Branchesi (Italy), Pierre Binetruy (France), Federico Ferrini (Italy), Richard N. Manchester (Australia), David H. Shoemaker (United States), Robin Stebbins (United States)

Commission E1–Solar Radiation and Structure

President:
Natalie Krivova
MPI für Sonnensystemforschung
Justus-von-Liebig-Weg 3
37077–Göttingen Germany

Vice-President:
Alexander Kosovichev
New Jersey Institute of Technology
Department of Physics
Tiernan Hall, Room 463
University Heights
07102–Newark
United States

Secretary:
Nicolas Labrosse
University of Glasgow
School of Physics and Astronomy
G12 8QQ –Glasgow
United Kingdom

Organizing Committee Members:

Michele Bianda (Switzerland), Nataliia Gennadievna Shchukina (Ukraine), Natalie Krivova (Germany), Alexander Kosovichev (United States), Nicolas Labrosse (United Kingdom), Gianna Cauzzi (Italy), Yoshinori SUEMATSU (Japan)

Commission E2–Solar Activity

President:
Lyndsay Fletcher
University of Glasgow
School of Physics and Astronomy
Kelvin Bldg
G12 8QQ–Glasgow
United Kingdom

Vice-President:
Paul S. Cally
Monash University
School of Mathematical Sciences
and Monash Centre for Astrophysics
Wellington Road
Clayton
VIC 3800–Victoria
Australia

Organizing Committee Members:

Manolis K. Georgoulis (Greece), Lyndsay Fletcher (United Kingdom), Karel J. Schrijver (United States), Paul S. Cally (Australia), Philippa K. Browning (United Kingdom), Jongchul CHAE (Korea, Rep of), Amy R. Winebarger (United States)

Commission E3–Solar Impact Throughout the Heliosphere

President:
Ingrid Mann
EISCAT Scientific Association
P. O. Box 812
98128–Kiruna
Sweden

Vice-President:
Carine Briand
Observatoire de Paris
LESIA/CNRS
5 place J. Janssen
92195–Meudon
France

Organizing Committee Members:

Ingrid Mann (Sweden), Margit Haberreiter (Switzerland), Olga Malandraki (Greece), Dibyendu Nandi (India), Kanya KUSANO (Japan), Ilya G. Usoskin (Finland), Carine Briand (France)

Commission F1–Meteors, Meteorites and Interplanetary Dust

President:
Jiří Borovička
Astronomical Institute of the Academy
of Sciences of the Czech Republic
Department of Interplanetary Matter
Fričova 298
251 65–Ondřejov
Czech Republic

Vice-President:
Diego Janches
Goddard Space Flight Center/NASA
Space Weather Lab
Code 674
20771–Greenbelt
United States

Secretary:
Pavel Koten
Astronomical Inst. Academy of Sciences
Department of Interplanetary Matter
Fričova 298
251 65–Ondřejov
Czech Republic

Organizing Committee Members:

Jiří Borovička (Czech Republic), Pavel Koten (Czech Republic), Margaret D. Campbell-Brown (Canada), Petrus Matheus Marie Jenniskens (United States), Galina O. Ryabova (Russian Federation), Jérémie J. Vaubaillon (France), David J. Asher (United Kingdom), Diego Janches (United States)

Commission F2–Exoplanets and the Solar system

President:
Alain Lecavelier des Etangs
Institut d'Astrophysique de Paris
98bis Bd Arago
75014–Paris
France

Vice-President:
Jack J. Lissauer
NASA/ARC
MS 245 3
CA 94035 1000–Moffett Field
United States

Organizing Committee Members:

Alain Lecavelier des Etangs (France), Régis Courtin (France), Mark T Lemmon (United States), Patrick Michel (France), Alessandro Morbidelli (France), Feng TIAN (China Nanjing), Paul A. Wiegert (Canada), Jack J. Lissauer (United States)

Commission F3–Astrobiology

President:
Sun Kwok
University of Hong Kong
Faculty of Science
Chong Yuet Ming Physics Bldg
Pokfulam Rd
Hong Kong
China Nanjing

Vice-President:
Masatoshi OHISHI
NAOJ
Astronomy Data Center
2-21-1 Osawa
Mitaka
181-8588–Tokyo
Japan

Secretary:
Joseph A. Nuth
NASA/GSFC
Solar System Exploration Division
Code 690
MD 20771–Greenbelt
United States

Organizing Committee Members:

Joseph A. Nuth (United States), Sun Kwok (China Nanjing), Masatoshi OHISHI (Japan), Muriel Gargaud (France), Jesus Martinez-Frias (Spain), Sergio Pilling (Brazil), Nils G. Holm (Sweden)

Commission G1–Binary and Multiple Star Systems

President:
Andrej Prša
Villanova University
Dept. of Astrophysics and Planetary Science
800 E Lancaster Ave
19085–Villanova
United States

Vice-President:
Virginia Trimble
University of California Irvine
Physics-Astronomy Department
4575 Physics
CA 92697-4575–Irvine
United States

Secretary:
Christopher Adam Tout
University of Cambridge
Institute of Astronomy
The Observatories
Madingley Road
CB3 0HA–Cambridge
United Kingdom

Organizing Committee Members:

Andrej Prša (United States), Virginia Trimble (United States), Brian D. Mason (United States), Robert D. Mathieu (United States), Terry D. Oswalt (United States), John Southworth (United Kingdom), Tomaž Zwitter (Slovenia, the Republic of), Christopher Adam Tout (United Kingdom)

Commission G2–Massive Stars

President:
Artemio Herrero Davó
Instituto de Astrofísica de Canarias
Research Division
C/ Vía Láctea s/n
38200–La Laguna, Tenerife
Spain

Vice-President:
Jorick S. Vink
Armagh Observatory
College Hill
BT61 9DG–Armagh
United Kingdom

Secretary:
Nicole St-Louis
Université de Montréal
Département de Physique
CP 6128, Succ Centre-Ville
QC H3C 3J7–Montréal
Canada

Organizing Committee Members:

You-Hua Chu (China Taipei), Artemio Herrero Davó (Spain), Jorick S. Vink (United Kingdom), Nicole St-Louis (Canada), Jose H. Groh (Switzerland), Gregor Rauw (Belgium), Asif ud-Doula (United States)

Commission G3–Stellar Evolution

President:
John C. Lattanzio
Monash University
School of Physics and Astronomy
10 College Walk
Clayton Campus
VIC 3800–Victoria
Australia

Vice-President:
Marc Howard Pinsonneault
Ohio State University
Astronomy Dept
140 W 18th Ave
OH 43210-1106–Columbus
United States

Secretary:
Jacco Th. van Loon
Keele University
Astrophysics Group
Lennard-Jones Laboratories
ST5 5BG–Staffordshire
United Kingdom

Organizing Committee Members:

Jacco Th. van Loon (United Kingdom), John C. Lattanzio (Australia), Marc Howard Pinsonneault (United States), Zhanwen HAN (China Nanjing), Franz Kerschbaum (Austria), Marco Limongi (Italy), Marcella Marconi (Italy), Gražina Tautvaišiene (Lithuania)

Commission G4–Pulsating Stars

President:
Christopher Simon Jeffery
Armagh Observatory
College Hill
BT61 9DG–Armagh
United Kingdom

Vice-President:
Jaymie Matthews
University of British Columbia
Dept of Physics and Astronomy
Hennings Bldg
6224 Agricultural Rd
BC V6T 1Z1–Vancouver
Canada

Secretary:
Saskia Hekker
Max Planck Institute for Solar System Research
Justus von Liebigweg 3
37077–Göttingen
Germany

Organizing Committee Members:

Christopher Simon Jeffery (United Kingdom), Jaymie Matthews (Canada), Saskia Hekker (Germany), Joyce Ann Guzik (United States), Karen Pollard (New Zealand), Hiromoto SHIBAHASHI (Japan), Dennis Stello (Australia)

Commission G5–Stellar and Planetary Atmospheres

President:
Ivan Hubeny
University of Arizona
Steward Observatory and Department of Astronomy
933 N. Cherry Avenue
AZ 85721–Tuscon
United States

Vice-President:
Carlos Allende Prieto
Instituto de Astrofisica de Canarias
Via Lactea s/n
38200–La Laguna, Tenerife
Spain

Organizing Committee Members:

Ivan Hubeny (United States), France Allard (France), Katia Cunha (Brazil), John D. Landstreet (Canada), Thierry M Lanz (France), Lyudmila I. Mashonkina (Russian Federation), Adam P. Showman (United States), Carlos Allende Prieto (Spain)

Commission H1–The Local Universe

President:
Eva K. Grebel
Astronomisches Rechen-Institut
Zentrum fuer Astronomie der Universitaet Heidelberg
Moenchhofstr. 12-14
University of Heidelberg
69120–Heidelberg
Germany

Vice-President:
Dante Minniti
Universidad Andres Bello
Department of Physics
Fernandez Concha 700
Las Condes
Santiago
Chile

Organizing Committee Members:

Eva K. Grebel (Germany), Dante Minniti (Chile), Evangelie (Lia) Athanassoula (France), Sofia Feltzing (Sweden), Yasuo FUKUI (Japan), Vanessa M. Hill (France), Margaret Meixner (United States), Gang ZHAO (China Nanjing)

Commission H2–Astrochemistry

President:
Thomas J. Millar
Queen's Univ Belfast
School of Maths-Physics
13 Stranmillis Rd
BT9 5FS–Belfast
United Kingdom

Vice-President:
Edwin A. Bergin
University of Michigan
Astronomy Dept
1085 S. University Ave
MI 48109-1107
Ann Arbor
United States

Secretary:
Maria R. Cunningham
UNSW
School of Physics
NSW 2052–Sydney
Australia

Organizing Committee Members:

Maria R. Cunningham (Australia), Thomas J. Millar (United Kingdom), Satoshi Yamamoto (Japan), Edwin A. Bergin (United States), Yuri Aikawa (Japan), Paola Caselli (Germany), Jes K. Jörgensen (Denmark)

Organizing Committee Members:

Eva K. Grebel (Germany), Dante Minniti (Chile), Evangelie (Lia) Athanassoula (France), Sofia Feltzing (Sweden), Yasuo FUKUI (Japan), Vanessa M. Hill (France), Margaret Meixner (United States), Gang ZHAO (China Nanjing)

Commission H3–Planetary Nebulae

President:
Letizia Stanghellini
NOAO
950 North Cherry Ave
AZ 85726-6732–Tucson
United States

Vice-President:
Albert Zijlstra
University of Manchester
School of Physics & Astronomy
Oxford Road
M13 9PL–Manchester
United Kingdom

Secretary:
Arturo Manchado
Instituto de Astrofísica de Canarias
Research
C/ Vía Láctea s/n
38200–La Laguna, Tenerife
Spain

Organizing Committee Members:

Letizia Stanghellini (United States), Arturo Manchado (Spain), Orsola De Marco (Australia), Karen B. Kwitter (United States), Miriam Pena (Mexico), Albert Zijlstra (United Kingdom)

Inter-Divisions Commissions (2015–2018)

Inter-Division A-F Commission–Celestial Mechanics and Dynamical Astronomy

President:
Cristian Beaugé
Observatorio Astronómico
Laprida 854
Barrio Observatorio
5000–Córdoba
Argentina

Vice-President:
Alessandra Celletti
Università di Roma Tor Vergata
Dipto di Matematica
Via della Ricerca Scientifica, 1
00133–Roma
Italy

Secretary:
Bonnie Alice Steves
Glasgow Caledonian University
The Graduate School
Cowcaddens Rd
G4 0BA–Glasgow
United Kingdom

Organizing Committee Members:

Cristian Beaugé (Argentina), Bonnie Alice Steves (United Kingdom), Douglas P. Hamilton (United States), Eiichiro KOKUBO (Japan), Jacques Laskar (France), Daniel J. Scheeres (United States), Alessandra Celletti (Italy)

Inter-Division B-C Commission–Protection of Existing and Potential Observatory Sites

President:
Richard F. Green
University of Arizona
Steward Observatory
933 North Cherry Ave
AZ 85721-0065 -Tucson United States

Vice-President:
Constance Elaine Walker
National Optical Astronomy Observatory
950 North Cherry Avenue
85719 -Tucson
United States

Organizing Committee Members:

Richard F. Green (United States), David Galadí-Enríquez (Spain), Harvey Steven Liszt (United States), Ramotholo R. Sefako (South Africa), Yongheng ZHAO (China Nanjing), Constance Elaine Walker (United States)

Inter-Divisions G H J Commission–Stellar Clusters throughout Cosmic Space and Time

President:
Richard de Grijs
Peking University
Kavli Institute for Astronomy and Astrophysics
Yi He Yuan Lu 5
Hai Dian District
100871–Beijing
China Nanjing

Vice-President:
Amanda I. Karakas
Monash University
Monash Centre for Astrophysics, School of Physics & Astronomy
School of Physics & Astronomy, Monash University
Clayton
3800–Melbourne
Australia

Secretary:
Ernst Paunzen
Faculty of Sciences Masaryk University
Department of Theoretical Physics and Astrophysics
Kotlářská 2
611 37–Brno
Czech Republic

Organizing Committee Members:

Ernst Paunzen (Czech Republic), Francesca D'Antona (Italy), Richard de Grijs (China Nanjing), Amanda I. Karakas (Australia), André Moitinho (Portugal), Jan Palouš (Czech Republic), Alison I. Sills (Canada)

Inter-Division D-G-H-J Commission–Galaxy Spectral Energy Distributions

President:
Denis Burgarella
Aix-Marseille University
Laboratoire d'Astrophysique de Marseille
Technopole de Chateau-Gombert
38, rue Joliot-Curie
13388–Marseille Cedex 13
France

Vice-President:
Cristina Carmen Popescu
University of Central Lancashire
Jeremiah Horrocks Institute
PR1 2HE –Preston
United Kingdom

Secretary:
Daniel Schaerer
Université de Genève
Observatoire de Genève
51 chemin des Maillettes
1290–Versoix
Switzerland

Organizing Committee Members:

Denis Burgarella (France), Cristina Carmen Popescu (United Kingdom), Amy J. Barger (United States), Asantha R. Cooray (United States), Robert C. Kennicutt (United Kingdom), Toru YAMADA (Japan), Daniel Schaerer (Switzerland)

Inter-Division B-H-J Commission–Intergalactic Medium

President:
Avery A. Meiksin
University of Edinburgh
Institute for Astronomy
Royal Observatory
Blackford Hill
EH9 3HJ–Edinburgh
United Kingdom

Vice-President:
Hsiao-Wen CHEN
Kavli Institute for Cosmological Physics
Department of Astronomy & Astrophysics
The University of Chicago
5640 S. Ellis Ave
Chicago, IL 60637
United States

Organizing Committee Members:

Avery A. Meiksin (United Kingdom), Hsiao-Wen CHEN (United States), Valentina D' Odorico (Italy), Nissim Kanekar (India), Jason X. Prochaska (United States), Joop Schaye (Netherlands)

Cross-Divisions Commissions (2015–2018)

Cross-Division D-J Commission–Supermassive Black Holes, Feedback and Galaxy Evolution

President:
William Richard Forman
Smithsonian Observatory
Harvard Smithsonian Center for Astrophysics
High Energy Astrophysics Division
60 Garden Street
MA 02138-1516–Cambridge
United States

Vice-President:
Thaisa Storchi-Bergmann
UFRGS
Inst de Física
Campus do Vale
CP 15051
91501-970–Porto Alegre, RS
Brazil

Organizing Committee Members:

William Richard Forman (United States), Thaisa Storchi-Bergmann (Brazil), Judith H. Croston (United Kingdom), Sebastian Heinz (United States), Roberto Maiolino (United Kingdom), Sera B. Markoff (Netherlands), Hagai Netzer (Israel), Marta Volonteri (France)

Cross-Division A-F Commission–Solar System Ephemerides

President:
Andrea Milani Comparetti
Università di Pisa
Dipartimento di Matematica
Piazzale B. Pontecorvo 5
56127–Pisa
Italy

Vice-President:
William M. Folkner
JPL m/s 301-121
4800 Oak Grove Dr.
91109–Pasadena
United States

Organizing Committee Members:

William M. Folkner (United States), Steven R. Chesley (United States), Andrea Milani Comparetti (Italy), Jean-Eudes Arlot (France), Elena V. Pitjeva (Russian Federation), Paolo Tanga (France)

Transactions IAU, Volume XXX
Proc. XXX IAU General Assembly, August 2018
Debra Elmegreen, ed.
doi:10.1017/S1743921323005690

CHAPTER II: TWENTY NINTH GENERAL ASSEMBLY BUSINESS SESSIONS – CORRIGENDUM

Norio Kaifu

doi:10.1017/S1743921318003952, Published online by Cambridge University Press, 29 January 2019

The Author and Cambridge University Press apologise for an error in the above article. In the article title, 'Twenty Ninth' was originally given as 'Twenty Nineth'.

This has been updated.

Reference

Kaifu, N. (2015). CHAPTER II: TWENTY NINTH GENERAL ASSEMBLY BUSINESS SESSIONS. *Proceedings of the International Astronomical Union*, 13(T29B), 12–15. doi:10.1017/S1743921318003952

Transactions IAU, Volume XXX
Proc. XXX IAU General Assembly, August 2018
Debra Elmegreen, ed.
doi:10.1017/S1743921323005707

CHAPTER II: TWENTY NINTH GENERAL ASSEMBLY BUSINESS SESSIONS : SECOND SESSION – CORRIGENDUM

Norio Kaifu

doi:10.1017/S1743921318003964, Published online by Cambridge University Press, 29 January 2019

The Author and Cambridge University Press apologise for an error in the above article. In the article title, 'Twenty Ninth' was originally given as 'Twenty Nineth'.

This has been updated.

Reference

Kaifu N. CHAPTER II: TWENTY NINTH GENERAL ASSEMBLY BUSINESS SESSIONS. *Proceedings of the International Astronomical Union.* 2015;13(T29B):16–22. doi:10.1017/S1743921318003964

Transactions IAU, Volume XXX
Proc. XXX IAU General Assembly, August 2018
Debra Elmegreen, ed.
doi:10.1017/S1743921323005719

CHAPTER III: TWENTY NINTH GENERAL ASSEMBLY CLOSING CEREMONY – CORRIGENDUM

doi:10.1017/S1743921318003976, Published online by Cambridge University Press, 29 January 2019

The Author and Cambridge University Press apologise for an error in the above article. In the article title, 'Twenty Ninth' was originally given as 'Twenty Nineth'.

This has been updated.

Reference

CHAPTER III: TWENTY NINTH GENERAL ASSEMBLY CLOSING CEREMONY. *Proceedings of the International Astronomical Union*. 2015;13(T29B):23–29. doi:10.1017/S1743921318003976

Transactions IAU, Volume XXX
Proc. XXX IAU General Assembly, August 2018
Debra Elmegreen, ed.
doi:10.1017/S1743921323005720

CHAPTER IV: TWENTY NINTH GENERAL ASSEMBLY RESOLUTIONS OF THE XXIXth GENERAL ASSEMBLY – CORRIGENDUM

doi:10.1017/S1743921318003988, Published online by Cambridge University Press, 29 January 2019

The Author and Cambridge University Press apologise for an error in the above article. In the article title, 'Twenty Ninth' was originally given as 'Twenty Ningth'.

This has been updated.

Reference

CHAPTER IV: TWENTY NINTH GENERAL ASSEMBLY. *Proceedings of the International Astronomical Union.* 2015;13(T29B):30–49. doi:10.1017/S1743921318003988

Transactions IAU, Volume XXX
Proc. XXX IAU General Assembly, August 2018
Debra Elmegreen, ed.
doi:10.1017/S1743921323005732

CHAPTER VII: TWENTY NINTH GENERAL ASSEMBLY: IAU STATUTES, BYE-LAWS, AND WORKING RULES – CORRIGENDUM

doi:10.1017/S1743921318004027, Published online by Cambridge University Press, 29 January 2019

The Author and Cambridge University Press apologise for an error in the above article. In the article title, 'Twenty Ninth' was originally given as 'Twenty Nineth'.

This has been updated.

Reference

CHAPTER VII: TWENTY NINTH GENERAL ASSEMBLY. *Proceedings of the International Astronomical Union*. 2015;13(T29B):86–120. doi:10.1017/S1743921318004027